ADVANCES IN BIOTECHNOLOGICAL PROCESSES
VOLUME 9

Biotechnology in Agriculture

ADVANCES IN BIOTECHNOLOGICAL PROCESSES

Series Editor

Avshalom Mizrahi
Israel Institute for Biological Research
Ness Ziona, Israel

TITLES IN THE SERIES

Advances in Biotechnological Processes, Volume 1
Avshalom Mizrahi and Antonius L. van Wezel, Editors

Advances in Biotechnological Processes, Volume 2
Avshalom Mizrahi and Antonius L. van Wezel, Editors

Advances in Biotechnological Processes, Volume 3
Avshalom Mizrahi and Antonius L. van Wezel, Editors

Advances in Biotechnological Processes, Volume 4
Avshalom Mizrahi and Antonius L. van Wezel, Editors

Advances in Biotechnological Processes, Volume 5
Avshalom Mizrahi and Antonius L. van Wezel, Editors

Advances in Biotechnological Processes, Volume 6
Avshalom Mizrahi, Editor

Advances in Biotechnological Processes, Volume 7
Upstream Processes: Equipment and Techniques
Avshalom Mizrahi, Editor

Advances in Biotechnological Processes, Volume 8
Downstream Processes: Equipment and Techniques
Avshalom Mizrahi, Editor

Advances in Biotechnological Processes, Volume 9
Biotechnology in Agriculture
Avshalom Mizrahi, Editor

Contents of Previous Volumes Appears After the Index

ADVANCES IN BIOTECHNOLOGICAL PROCESSES
VOLUME 9

Biotechnology in Agriculture

Editor
Avshalom Mizrahi

Israel Institute for Biological Research
Ness Ziona, Israel

Alan R. Liss, Inc., New York

Address all Inquiries to the Publisher
Alan R. Liss, Inc., 41 East 11th Street, New York, NY 10003

LC 83-644876
ISBN 0-8451-3208-3

Contents

vi ◇ Contents

Somatic Embryogenesis and Polyembryogenesis in Conifers
D.J. Durzan and Pramod K. Gupta

Virus Detection in Squash-Blots of Plants and Insects: Applications in Diagnostics, Epidemiology, and Breeding
Henryk Czosnek and Nir Navot

Alkaloid Production by Plant Cell Cultures
C.A. Hay, L.A. Anderson, M.F. Roberts, and J.D. Phillipson

Potential for Exploiting Vesicular-Arbuscular Mycorrhizas in Agriculture
I.R. Hall

Electric Gene Transfer Into Plant Protoplasts and Cells
Hiromichi Morikawa, Asako Iida, and Yasuyuki Yamada

Applications of Somaclonal Variation
David A. Evans

Encapsulated Plant Embryos
Keith Redenbaugh, Jo Ann Fujii, and David Slade

Contributors

L.A. Anderson, Department of Pharmacognosy, The School of Pharmacy, University of London, London WC1N 1AX, England **[97]**

Henryk Czosnek, Department of Field and Vegetable Crops, Faculty of Agriculture, The Hebrew University of Jerusalem, Rehovot 76100, Israel **[83]**

D.J. Durzan, Department of Environmental Horticulture, University of California, Davis, CA 95616 **[53]**

David A. Evans, DNA Plant Technology Corporation, Cinnaminson, NJ 08077 **[203]**

Jo Ann Fujii, Plant Genetics, Inc., Davis, CA 95616 **[225]**

Andrew Goldsworthy, Department of Pure and Applied Biology, Imperial College, London SW7 2BB, England **[35]**

Pramod K. Gupta, Department of Environmental Horticulture, University of California, Davis, CA 95616 **[53]**

I.R. Hall, Invermay Agricultural Centre, Ministry of Agriculture and Fisheries, Mosgiel, New Zealand **[141]**

C.A. Hay, Department of Pharmacognosy, The School of Pharmacy, University of London, London WC1N 1AX, England **[97]**

Asako Iida, Research Center for Cell and Tissue Culture, Faculty of Agriculture, Kyoto University, Kyoto 606, Japan **[175]**

Hiromichi Morikawa, Research Center for Cell and Tissue Culture, Faculty of Agriculture, Kyoto University, Kyoto 606, Japan **[175]**

Nir Navot, Department of Field and Vegetable Crops, Faculty of Agriculture, The Hebrew University of Jerusalem, Rehovot 76100, Israel **[83]**

A.J. Parr, Plant Cell Culture Group, AFRC Institute of Food Research, Norwich Laboratory, Norwich NR4 7UA, England **[1]**

J.D. Phillipson, Department of Pharmacognosy, The School of Pharmacy, University of London, London WC1N 1AX, England **[97]**

Keith Redenbaugh, Plant Genetics, Inc., Davis, CA 95616 **[225]**

M.F. Roberts, Department of Pharmacognosy, The School of Pharmacy, University of London, London WC1N 1AX, England **[97]**

David Slade, Plant Genetics, Inc., Davis, CA 95616 **[225]**

Yasuyuki Yamada, Research Center for Cell and Tissue Culture, Faculty of Agriculture, Kyoto University, Kyoto 606, Japan **[175]**

Preface

The title *Biotechnology in Agriculture* suggests a broad area of scientific investigation. In fact, this volume of *Advances in Biotechnological Processes* covers aspects of plant technology designed to improve the quality, consistency, and productivity of cash crops.

Two of the nine chapters examine the derivation and possible applications of secondary products from cell culture, among them the alkaloids used in pharmaceutical preparations and the flavonoids used in the flavor and fragrance industries. Other chapters are devoted to somaclonal variation and growth control as examined through creation and analysis of plant tissue cultures. Laboratory management of plant growth stimulation by electrical treatment is also discussed in some depth, as are strategies for implementing somatic embryogenesis as a means of enhancing the gene pool.

Somatic hybridization by electric gene transfer is also reviewed, as is the range of processes available for production of artificial "seeds" to increase crop uniformity and yield. Another aspect of biotechnology and agriculture presented here is the use of vesicular-arbuscular mycorrhizas (VAM)-inocula to improve germination and yield of such significant food and feed crops as alfalfa, soybeans, potatoes, clover, fava beans, and apples, as well as other fruits. Virus detection with a squash-blot technique is detailed in terms of diagnostics, including availability and ease of application; epidemiology; and breeding.

This volume thus presents a wide range of interesting topics of interest to scientists and students of plant technology, agricultural research, and crop development worldwide.

Avshalom Mizrahi

Biotechnology in Agriculture, pages 1–34
© **1988 Alan R. Liss, Inc.**

Secondary Products From Plant Cell Culture

A.J. Parr

Plant Cell Culture Group, AFRC Institute of Food Research, Norwich Laboratory, Norwich NR4 7UA, England

Plant secondary products are compounds present in plants which are believed to have no role in the basic life process but, rather, have secondary, nonessential roles. They are characterized by their extreme chemical diversity; ranging from, for example, simple amines and esters of molecular weight less than 100, through the complex polynuclear heterocyclic compounds and conjugates with molecular weights in excess of 1000. The range of minor structural modifications between related compounds is also large—for example, approximately 1,000 related indole alkaloids have so far been isolated from plants [1]. This is not the place to enter into discussion of the possible roles of secondary products in plants [for reviews, see 2,3]; however, it seems likely that many are ecologically important in interactions between the plant and other organisms. The commercial importance of secondary products lies in the fact that many are active in man. Some have characteristic odors (e.g., the essential oils), and are used in the cosmetic industry; others possess characteristic flavors, and are used in the food industry (e.g., the bittering agent quinine). The most important groups are those which are pharmacologically active and which thus find uses as drugs. It has been estimated that in 1980, some 25% of drugs prescribed in the United States were either of plant origin or derived from precursors isolated from plants [4]. At present desirable secondary products are isolated from plants grown on a large scale in commercial plantations, or on a small scale by "cottage industries." This has a number of associated problems, including constraints on the quantities which can be produced, and susceptibility of production to environmental factors such as drought and pestilence. With the observation that plant cells can be cultured in vitro, and that these cultures can produce compounds typical of intact plants, there has for some years been an interest in exploiting plant cell cultures as an alternative source of secondary products. Although several problems remain to be solved before large-scale commercial exploitation becomes routine, the recent Japanese development of a commercial system for the production of the red pigment and antibacterial agent shikonin [5] gives cause for optimism. The present article gives an overview of the progress made, as of early 1986, in the use of plant cell cultures as an alternative to conventional agriculture in the production of plant secondary products. Since much of the early work has been extensively reviewed elsewhere, no attempt has been made to quote every reference available.

I. CELL CULTURES AND SYNTHESIS OF SECONDARY PRODUCTS

The process of taking plant cells into culture generally involves the induction of undifferentiated wound growth on sterile explants from the desired plant, under the influence of a combination of auxins (e.g., 2,4-dichlorophenoxyacetic acid = 2,4D) and cytokinins (e.g., kinetin, benzyl-adenine, or zeatin). The callus material is then removed from the explant and either maintained as such on solid medium supplemented with appropriate organic and inorganic nutrients [6–9] plus phytohormones, or else transferred into liquid suspension. The coordinated gene expression associated with intact plants is sacrificed in favor of the ease of growth and maintenance of biomass which is associated with undifferentiated cell cultures. The key problem in the exploitation of such cultures is thus the ability to activate and coordinate the required pathways so that the desired products are synthesized and accumulated in useful quantities. The synthetic capacity of many unmanipulated cultures is often quite low, with the more organized callus cultures frequently showing better productivity than dispersed cell suspensions [10]. Products which are synthesized in intact plants only in specific organs, and at specific times of the year—for example, humulone produced by maturing cones of hops, *Humulus lupulus*—may be produced at only very low levels, if at all, in unmanipulated cell cultures [11]. Products with less restricted synthesis in intact plants seem to be more readily produced in culture, possibly because a wider range of conditions is suitable for expression of the relevant pathways. Even here there may, however, be a tendency for productivity to decline with time, as, for example, seen by Sasse et al. [12] for β-carboline alkaloid synthesis in *Peganum harmala* callus cultures, and by Dhoot and Henshaw for hyoscyamine and scopolamine synthesis in *Hyoscyamus niger* suspension cultures [13]. Such effects may relate to the known genetic instability of many cell cultures [14,15]. Since strategies exist for improving the productivity of plant cell cultures (see sections II and III), the important thing to note, however, is that secondary product synthesis can indeed proceed in culture. The range of compounds so far detected is extensive and is summarized in Table I. It should be noted in passing that although synthesis occurs in culture, it may not proceed at all stages of the growth cycle. Some products, e.g., sinapyl alcohol in *Nicotiana tabacum* [16], are synthesized only by actively growing cultures in such a manner than product concentration per unit biomass always remains constant. These are termed growth-related products. Other products show non-growth-related synthesis. Thus, for instance, tripdiolide in *Tripterygium wilfordii* is synthesized only once growth slows towards the end of a growth cycle [17], while anthocyanins in *Populus* show a period of major synthesis during the lag phase following cell subculturing [18]. Some products show a combination of both growth-related and non-growth-related synthesis, the balance altering depending upon growth conditions [19].

As well as the question of levels of synthesis, two other factors related to

TABLE I. Some Secondary Products Which Have Been Detected in Plant Cell Cultures

Acridone alkaloids	Lignin
Alkaloids	Lipids/fatty acids
Alkanes	Methylxanthines
Alkanols and alkanones	Monoterpenoids
Amines	Napthoquinones
Amino acids	Opines
Anthocyanins	Peptides
Anthraquinones	Phenolics
Benzoic acid derivatives	Phthalides
Benzoquinones	Proteins
Betalains	Pterocarpans
Carbohydrates	Pyrethrins
β-Carboline alkaloids	Pyridine alkaloids
Cardenolides	Quinoline alkaloids
Carotenoids	Quinolizidine alkaloids
Chalcones	Retrochalcones
Cinnamic acid and derivatives	Rotenoids
Coumarins	Sapogenins/Saponins
Dianthrones	Sesquiterpenoids
Diterpenoids	Steroidal alkaloids
Flavones	Steroids
Flavonols	Tannins
Furanochromones	Terpenoids
Furanocoumarins	Thiarubrines
Glucosinolates	Thiophenes
Indole alkaloids	Triterpenoids
Iridoids	Tropane alkaloids
Isoquinoline alkaloids	Tropolones
Lignans	Valepotriates

control and coordination of gene expression are relevant to the exploitation of plant cells. First, the pattern of secondary products produced in culture need not resemble the pattern observed in intact plants [20,21]. This reflects both imbalances in metabolic expression in culture, as well as the absence of organ-specific synthesis or long-distance transport effects seen in plants. Where the item of interest is actually a complex mixture of related compounds, e.g., many essential oils and essences, then there may thus be problems in obtaining the desired composition. Secondly, although large sections of particular biosynthetic pathways may be expressed in culture, crucial enzymes in the pathway may be poorly expressed, and thus certain products may not always be synthesized. For instance, there is as yet no report of the synthesis of the important anticancer agent vinblastine in cell cultures of *Catharanthus roseus,* although it is produced in the intact plant. This possibly arises because present cell cultures can only readily synthesize one of the two related precursor alkaloids (catharanthine) from which vinblastine is composed [22]. The position of blocks in metabolism in culture has been found to be highly variable, and different cell lines can be isolated which show significant differences in the products which they accumulate [23]. The somewhat unpredictable nature of secondary metabolism in culture is often an important problem which has to be faced; one positive aspect of the subtle disturbances in metabolism is, however, that cell cultures may sometimes accumulate novel compounds not found in intact plants [24,25].

An alternative approach to the production of secondary products which may suffer from fewer of the problems described above is the use of organized or differentiated cultures. Under certain environmental conditions callus cultures often show a tendency to differentiate organs such as roots or shoots [26–28], or else special cell types such as laticifers [29]. This is frequently associated with improved production of secondary products normally associated with these structures in the intact plant. For example, undifferentiated callus of *Digitalis purpurea* produces little digitoxin, but shoot-forming callus can accumulate substantial quantities of this and other cardenolides [28]. The problem with callus cultures is that they are often difficult to grow and exploit on a large scale. The concept of organization and differentiation can, though, also be applied to suspension cultures. Some cultures routinely grow in such a way that newly divided cells do not separate, and, rather than a fine suspension, the culture consists of dense cell aggregates which can be up to several centimetres in diameter in some cases [30]. These aggregates occasionally show pronounced differentiation and organogenesis [e.g., 31]. Even when signs of differentiation are less pronounced, the organization inherent in aggregates might still produce beneficial effects on productivity. Thus there will be cell-to-cell contact, which allows intercellular communication via plasmodesmata, as in intact plants. Diffusion barriers also result in the establishment of gradients of O_2, CO_2, nutrients, and hormones. These factors might be important in influencing cell metabolism and, hence, secondary product synthesis [32].

One problem with the use of aggregated, organized cell cultures is that aggregate size is often difficult to manipulate and control. Also, not all cell lines will spontaneously form such cultures. Cells may, however, be obtained as what are essentially aggregates of defined size by the techniques of cell immobilization. Two basic approaches have developed in recent years [32–35]. In one, cells are mixed with a solution of a gelling polysaccharide or similar material [35], and then droplets or thin layers are induced to gel by the appropriate treatment, as detailed in Table II. Cells can then be grown on,

TABLE II. Gel Matrices for Cell Immobilization

Gelling agent	Gelling procedure
Alginate	Add 50–100 mM Ca^{+2}
Agarose or agar	Cool below gelling temperature
K-carrageenan	Cool below gelling temperature; 300 mM K^+ treatment stabilizes gel
Gelatin	Cool below gelling temperature; glutaraldehyde stabilizes gel but may damage cells
Polyacrylamide or polyacrylamide/polysaccharide mixture	Free-radical generation, e.g., persulfate, UV light

though too extensive growth results in disruption of the gel structure. The second approach to cell immobilization is to add particles of reticulated foam [36,37] or nylon [38] to a fine cell suspension. Cells become immobilized on the matrix, perhaps by means of an adhesive mucilage [37]. From a low initial innoculum, cells can then be grown up to fill the entire matrix volume [39,40]. The size of the support particles and the dimensions of their internal spaces can be optimized to suit the culture available, and the conditions under which the cells are to be grown [41,42]. Many situations have been described where the productivity of immobilized cells appears to exceed that of free cell suspensions [35,36,43]. Advantages presumably stem from the same factors of microenvironment and cell-to-cell communication descibed for naturally aggregated cultures. If cell-to-cell contact is an important factor, then it may be that growth of cells in reticulated foam is the most useful immobilization technique. Since cells grow up within the foam from an initial low-density innoculum to a final high cell density, interactions will be maximized.

Finally, when discussing organized cultures, reference should be made to fully differentiated organ cultures. These have until recently attracted little interest, since root and shoot cultures are often slow growing, tricky to maintain, and inappropriate to large-scale fermentation techology. Within the last year or two there have, however, been some exciting breakthroughs. Plant tissues, when infected with the bacterium *Agrobacterium rhizogenes,* produce a profuse outgrowth of roots from the point of infection. These roots have been genetically modified by the insertion of a fragment of DNA from a plasmid in the infecting bacterium [44]. The roots can be excised from the plant and cultured in liquid medium; no external hormones are necessary, since the transformed roots now produce their own, and growth is often extremely rapid. It has been shown that such roots produce secondary metabolites which are typically produced in the roots of intact plants, and with a similar level and pattern of products [45,46]. Since the transformed roots are robust and fast growing they can be exploited in fermenter systems, as shown by Rhodes et al. [47]. It is likely that *A. rhizogenes*-transformed roots will prove to be an exceptionally useful system, offering a solution to many of the problems associated with the use of conventional tissue cultures for the production of secondary products. The approach is at present limited to target compounds synthesized in roots; the use of mutants of the related *A. tumefaciens* to induce shoot cultures [48,49] might, however, be one approach to obtaining a wider variety of products.

II. IMPROVING PRODUCTIVITY BY MANIPULATING CELLULAR ENVIRONMENTS

One of the key problems with plant cell cultures as a source of fine chemicals is that the cultures, as first established, usually show a productivity which is well below that needed to make any exploitation economically feasible. Much effort has thus, in recent years, been put into the development of methods to enhance the synthesis of secondary products by plant cells. The simplest general approach to improving product yield is to manipulate the physical and chemical environment in which the cells are growing, in search of a more favorable set of conditions.

A number of parameters have been identified which can potentially influence productivity. These include—

- Hormonal regime
- Nutrient regime
- Presence of specific elicitor compounds
- pH of the growth medium
- Levels of O_2 and CO_2 and other gases
- Temperature
- Light quality and quantity

Interactions between these parameters might also be expected to be important, though the complex experiments necessary to look at these interactions in detail mean that, at present, information concerning this is of limited extent.

A. Hormone Regime

Except in the case of *Agrobacterium*-transformed tissue, which synthesizes enhanced levels of endogenous auxins, the establishment of cell cultures usually involves exposing cells to high levels of 2,4D or other potent auxins. Such conditions, while beneficial for cell growth, are with a few

TABLE III. The Effects of a Range of Auxins on Quinoline Alkaloid Synthesis in *Cinchona ledgeriana* Suspension Cultures*

Auxin (mg/liter)	Total alkaloid produced (μg/g fresh weight cells)
2,4-dichlorophenoxyacetic acid (0.5)	56
4-chlorophenoxyacetic acid (0.5)	72
2-methyl-4-chlorophenoxyacetic acid (0.5)	49
Napthaleneacetic acid (2.0)	103
Indoleacetic acid (2.0)	262

*Cells were grown as described by Rhodes et al. [55] in medium supplemented with 0.1 mg/liter zeatin riboside; then after 28 days the total alkaloid present (cells plus medium) was determined.

exceptions generally deleterious to secondary product formation. (One such exception is L-DOPA production by *Mucuna pruriens* [50].) Lindsey and Yeoman [51] and others have noted that an inverse correlation between growth rate and levels of secondary product synthesis tends to be a general feature of many cell cultures. Since the continued presence of auxins is necessary for culture maintenance and suppression of redifferentiation, it is usually impractical to remove auxins from the medium totally. The yield of secondary products can, however, be optimized by careful adjustment of the auxin level. In *Cinchona ledgeriana* suspension cultures, for example, Robins et al. [52] found alkaloid production to be 50-fold greater at 0.1 mg/liter 2,4D than at 2.0 mg/liter. At the low auxin level tissue regeneration eventually began to occur, and in the end a level of 0.5 mg/liter 2,4D was chosen as giving the best sustainable productivity [52]. In addition to the effects of auxin levels, the type of auxin used can also exert a powerful influence on secondary product formation (see, for example, Zenk et al. [53], and Table III). In general 2,4D is less good for product synthesis than the natural auxin indoleacetic acid (IAA) [54], with napthyleneacetic acid (NAA) tending to have an intermediate effect [e.g., 19,55].

As well as the effects of auxins, the levels and types of cytokinins present in the growth medium may additionally influence secondary product formation [e.g., 55]. When optimizing the production of any compound from a cell culture, a wide variety of media of differing hormonal level and composition is thus generally examined to find the best production medium. Since such a production medium often supports only poor growth, although the productivity per unit weight of cells may be quite good, the productivity per day can still be rather low. In these circumstances a two-stage approach can be used. In this, cells are initially grown in a specialized growth medium so that a large biomass is rapidly generated. The cells are then transferred to the production medium, where a high cell mass is now already available for product synthesis.

Most of the hormonal manipulations carried out with cell cultures are generally done on the assumption that cells will respond with a relatively long-term, stable adaptation to the new hormone treatment. It should not, however, be forgotten that the transient hormonal balance achieved during a switch from one regime to another can also be a powerful modulator of secondary product synthesis. Rhodes et al. [55] observed a transitory stimulation of alkaloid synthesis when cells of *C. ledgeriana* were transferred from a medium containing 2,4D and benzyladenine to one containing IAA and zeatin riboside. In such situations it may be necessary to cycle continually between media with different hormones in order to achieve the maximum production of secondary products [56].

Finally, it should be noted that there are certain cultures—e.g. spontaneously "habituated" lines [57] and *Agrobacterium tumefaciens*-transformed cultures—which can be grown as callus or in suspension without the need for exogenous hormones. This is because they synthesize enough endogenously to supply their needs. As might be expected, some of these cultures are known to be relatively little affected by external hormone combinations which promote secondary product formation in related hormone-dependent lines [58]. A comprehensive analysis of the effects of hormone independence on secondary product synthesis has yet to be made, though both instances where production is apparently decreased [59], and somewhat increased [58,60] have been noted.

B. Nutrient Regime

Plant cell cultures are usually grown heterotrophically, using a simple sugar as carbon source and an inorganic supply of other nutrients. In general sucrose appears better than glucose or fructose for secondary product formation [61]. The level of sugar supplied can also influence secondary metabolism [e.g., 60,62], but often in a rather unpredictable manner. Thus, high sucrose concentrations (approx. 10% w/v) stimulate anthraquinone synthesis in many species of the genus *Galium* but inhibit anthraquinone formation in others—e.g., *G. rubioides* [62]. Similarly, high sucrose levels (6–8%) can be used to stimulate alkaloid synthesis in some systems [60], while in others the optimum sucrose level is much lower, at around 2% [58].

The effects of inorganic nutrient supply have perhaps been less thoroughly investigated than the effect of sugars. It is, however, clear that different nutrient regimes can greatly influence the accumulation of secondary products. Thus, in common with some lower organisms [63], many plant cell cultures seem sensitive to the phosphate supply. Maintaining cellular phosphate levels below the optimum for growth has been found to stimulate cinnamoylputresine accumulation in *Nicotiana tabacum* cell suspension cultures by 3–4-fold [64,65], and low phosphate media also stimulate alkaloid production in *Catharanthus roseus* [66] and *Peganum harmala* [12]. The effects of other nutrients seem less predictable, but substantial changes in product levels can also be produced. Anthraquinone formation in *Morinda citrifolia* is, for example, sensitive to the level of nitrate in the medium— being optimal at 25–40 mM [53]. Although studies of the effects of other nutrients are very few, it is likely that significant effects will also be seen in some cases. The stimulation of production of certain phenolic compounds during boron deficiency [67] is one such example. Clearly, taking all aspects together, it is apparent that mineral nutrition can greatly affect not only cell growth, but also the specific rate of secondary product formation. In many cases the effects are, however, not immediately predictable, and optimum conditions have to be determined experimentally.

C. Elicitors

A number of plant products—the phytoalexins and related compounds— have antifungal or antibacterial properties, and are produced in response to challenge by potential pathogens. The compounds come from a very wide range of chemical families (see Fig. 1), with each plant species producing a characteristic spectrum of products. Their synthesis is known to be triggered by specific elicitor molecules, and this system is active in cell cultures as well as in intact plants. Thus, the synthesis of many of the appropriate phytoalexins can be induced by treating a culture with a purified elicitor or, since these are not often available, more commonly with a culture filtrate or autoclaved culture of a suitable fungus or other organism [68,69]. In some instances various abiotic agents such as polylysine and heavy metal salts also have elicitor activity [68,69]. The mode of action of elicitors is as yet still uncertain, but there is evidence that they may induce the synthesis of specific mRNAs [70,71].

Since many compounds have phytoalexin-like properties, the use of fungal or abiotic elicitors to stimulate product synthesis may be relevant to a number of areas. As well as the typical phytoalexins which are found in any quantity only in elicitor-treated cells, certain compounds which are more generally synthesized in tissue culture may also increase following elicitor treatment. Anthraquinone levels in *Cinchona ledgeriana* suspension cultures are, for example, doubled following treatment with autoclaved mycelia of *Phytophthora cinnamomi* [72], and there is also a greater degree of product release into the medium (cf. section IV.B). Levels of diosgenin in suspension cultures of *Dioscorea deltoidea* are similarly increased (by 70% in this case)

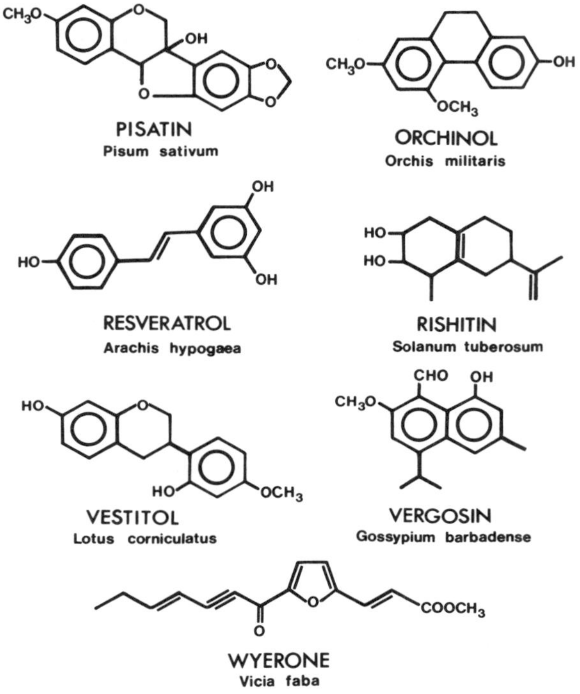

Fig. 1. Typical phytoalexins, and the species in which they are produced. (Not all of these compounds have as yet been studied in cell or tissue cultures.)

following treatment with autoclaved mycelia of *Rhizopus arrhizus* [73]. It must be remembered, of course, that not all compounds are capable of being induced by elicitors. Even where induction is possible, different elicitors may also show different efficacy [74].

D. Supply of Precursors

Under some conditions the flux through a particular biosynthetic pathway may be limited by the availability of precursors or particular metabolic intermediates. Supplying these compounds externally might thus increase the rate of product synthesis, provided of course they are taken up and reach the site of product biosynthesis. Lindsey and Yeoman [75] have had good success in stimulating capsaicin synthesis in some lines of *Capsicum frutescens* by supplying phenylalanine and isocapric acid, and Mérillon et al.

[60] have been able to stimulate ajmalicine production in a line of *Catharanthus roseus* approximately tenfold by supplying secologanin. Hay et al. [31] were also able to stimulate quinine formation in *Cinchona ledgeriana* fivefold by supplying 2.5 mM tryptophan. In some other studies, particularly where very early precursors are employed, stimulation of product formation is, however, small, or even nonexistent. Deus and Zenk [76], for example, found tryptophan to stimulate alkaloid production in one cell line of *C. roseus* by a factor of three, yet in two other cell lines there was either no effect or a partial inhibition of alkaloid synthesis. It would appear that in a significant number of cases the supply of particular precursors may not be the major rate-limiting factor in secondary product biosynthesis. When one considers that the high levels of precursors often required to have any action may also produce toxic side effects [e.g., 77], then it becomes clear that precursor feeding, while in many cases an extremely useful technique, need not necessarily be universally applicable.

Before moving on to consider other factors, a passing mention should be made to the concept of biotransformation, which in many ways is allied to precursor feeding. As with microorganisms, plant cell cultures are able to perform various regio- and stereospecific reactions on organic compounds added to their growth media. The possibility therefore exists to use cell cultures to upgrade a particular substance into another with a higher commercial or scientific value. Biotransformation differs from precursor feeding in that the transformations of interest are chemically simple, typically only a single step. The substrate for transformation also need not necessarily be a normal metabolite of the culture used. Interconvertion of steroids, and particularly the C-12 hydroxylation of cardiac glycosides, is one area which because of its medical relevance has received considerable attention in recent years [78,79]. As with de novo synthesis of plant secondary products, the ability of cultures to carry out biotransformations seems to be influenced by a variety of parameters—both genetic and environmental [78,79]. For a detailed examination of the biotransformation capacity of plant cell cultures, the reader is referred to the review of Reinhard and Alfermann [78].

E. Medium pH

The effect of pH on the production of secondary metabolites is a little-studied area, the culture medium's pH generally not being tightly controlled during growth. Instead the cells are often left to modify the medium pH by their own metabolic activity [80]. Even when tight control is not exerted over medium pH, there can, however, still be effects of the initial medium pH on secondary product accumulation. Thus Koul et al. [27] found an initial medium pH of 3.5 to result in a sevenfold greater production of alkaloid in *Hyoscyamus muticus* cultures than an initial pH of 6.0, despite the fact that the final pH of the two systems was very similar.

F. Levels of Oxygen, Carbon Dioxide, etc.

The role of aeration, O_2 supply, and CO_2 removal is another little-studied area, since these factors are difficult to study in small-scale systems. The discovery that the characteristics of many cultures alter following scale-up from shake flasks to fermenters where the gaseous exchange characteristics have been altered has, however, shown just how important these factors are [81,82]. Active work in the area is now taking place. Scragg et al. found that in one line of *Catharanthus roseus* grown under fermenter conditions in certain media, the addition of 4% CO_2 to the sparging gas was necessary to retain alkaloid productivity [82]. In these fermenter runs an oxygen saturation of $dO_2 = 60\%$ was also considerably better for alkaloid production than $dO_2 = 20\%$ [82]. Breuling et al. [83] similarly found higher oxygen tensions to be beneficial for berberine and jatrorrhizine production by cells of *Berberis wilsonae* grown in airlift fermenters. On the other hand there is circumstantial evidence that capsaicin accumulation in *Capsicum frutescens* might actually be enhanced by low oxygen tensions. Thus, a transitory loss of air supply to an immobilized cell fermenter was followed by the appearance of a burst of capsaicin [42]. Clearly, the effects of gas supply on secondary product synthesis are complex, and much further investigation is needed.

G. Temperature

This is yet another poorly explored area. In the few systems that have been studied, alkaloid production per gram of tissue was found to be optimal at 25°C in both *Nicotiana* [84] and *Peganum* [85]. In *Nicotiana,* production at 20°C and at 30°C was only about half the maximum value. The exact significance of the precise optimum temperature is uncertain, since stocks of the lines used were likely to have been maintained at only a single temperature, and thus the cells were presumably preconditioned to a certain temperature regime. The importance of temperature in controlling secondary product accumulation in some lines is, however, clear.

H. Light

Although most plant cell cultures are not fully photosynthetically competent and are rarely exposed to photosynthetically significant light intensities, the behavior of cultures can be influenced by photoperiodicity, light quality, and light intensity [86]. Many instances of light effects on secondary metabolism are known, and some of these systems are quite well understood, even at the molecular level. In *Petroselinum hortense* flavonoid biosynthesis is greatly stimulated by UV light, and flavone levels may increase roughly 40-fold [87]. The mechanism apparently involves a phytochrome-mediated induction of specific mRNA synthesis which results in a transient, but dramatic, rise in the levels of various flavonoid biosynthetic enzymes such as phenylalanine ammonia-lyase and chalcone synthase [70,87]. Somewhat similar events may also be involved in the blue-light-induced synthesis of anthocyanins in *Haplopappus gracilis* [88] and *Populus* cultures [18], and the less well characterized light enhancement of polyphenolic accumulation in *Rosa* sp. [89] could similarly be a further example of this type of process.

Light stimulation of secondary product synthesis or accumulation is not limited to phenolic compounds. The accumulation of some steroidal products such as diosgenin in *Dioscorea* and solasodine in *Solanum* can be enhanced by white light [86], and so too can the accumulation of certain alkaloids, e.g., serpentine in some lines of *Catharanthus roseus* [90]. Similarly the pattern of products produced by a culture is open to manipulation by light in some systems. The pattern of essential oil components produced by a culture of *Ruta graveolens* was, for instance, significantly altered when the cells were exposed to blue light as opposed to either red light or darkness, although oil production actually occurred in all instances [91,92].

Although the best characterized examples of the effects of light, as discussed above, involve a stimulation of secondary product accumulation, it should be noted that there are also systems where light has been found to have an inhibitory effect. For example, in *Lithospermum erythrorhizon* the formation of shikonin derivatives is inhibited by blue or white light [93], and light suppression of alkaloid accumulation has been observed in *Scopolia parviflora* [26], *Nicotiana tabacum* [94], and *Cinchona ledgeriana* [58]. There has been speculation that some of these effects may be due to photodegradation of products [95], but this cannot be the full story. In *C. ledgeriana*, one cell line was isolated in which alkaloid production was slightly ($\times$ 1.5) stimulated by light, and another line transformed with the bacterium *Agrobacterium tumefaciens* was obtained in which alkaloid production was essentially unaffected by light (Payne, Robins, and Rhodes, unpublished results). A second *A. tumefaciens*-transformed line however showed a very substantial "light inhibition" of alkaloid accumulation, with levels in the dark sometimes being as high as 50 times the level under 600-lux white light [58]. Investigation of enzyme levels in dark adapted cells revealed a 5–10-fold higher level of tryptophan decarboxylase and strictosidine synthase than in light-grown cells [96]. These enzyme changes could be related to the enhanced alkaloid accumulation seen.

III. GENETIC VARIATION AS A SOURCE OF ENHANCED PRODUCTIVITY

Improvements of productivity arising from manipulation of the culture environment generally have to be obtained empirically. The cells are exposed to a series of conditions without much prior knowledge as to which treatments, if any, will have an effect on productivity, and whether these effects will be beneficial or deleterious. Although the scope for culture improvement by these methods is enormous, it would also be useful to have methods with more predictable success and which can exploit other controlling factors in the regulation of secondary product metabolism. Such methods are available, and are based on the inherent genetic variability of plants and on the enhancement of this variation under some tissue culture conditions, as seen in the phenomenon of somaclonal variation [15,97].

A typical starting point in the genetic approach to enhancement of productivity is to screen individual intact plants for their ability to accumulate products, before inducing cell cultures from them. The rationale is that high-producing plants should give rise to high-producing cultures. Averaging over all cell isolates obtained from a particular plant this does, in general, seem to hold true—as, for example, shown by Zenk et al. [19] for serpentine production in *Catharanthus roseus,* and Kinnersley and Dougall [98] for nicotine production in *Nicotiana tabacum.* At the level of individual cultures the position is, however, less clear-cut; and there are several examples of lines from high-producing parent plants showing lower productivity than certain lines derived from low-producing plants [12,19,90]. When one additionally considers that high levels of production, even if achieved, may be unstable with time [10,12], then it is clear that the initial screening approach may sometimes have limitations. It is, however, a useful starting point for further improvements. Having made this qualification, recent developments suggest there may, though, be one particular area where screening of intact plants could be of very major importance. This is in the field of *Agrobacterium rhizogenes*-transformed root cultures. There is growing evidence that such root cultures produce the same spectrum of products as is produced in the roots of the parent plant, and at a level similar to, or even slightly above, that found in the plant [45]. Such production is also stable with time [46]. The transformed root system thus seems to offer a reliable way of obtaining the precise biosynthetic potential of the parent plant in a biotechnologically amenable form.

Once a suitable callus or suspension culture has been obtained, further improvement is possible by taking advantage of the fact that all cells in the culture will not be identical. This variation is derived from a low frequency (approximately 1 in 10^6) natural mutation rate, which can be enhanced by artificial mutagenesis if necessary, or from the much higher frequency (even up to 1 in 10 in extreme cases) culture-induced variation seen in many cell cultures. The causes of culture-induced variation are still under dispute, but much seems to arise from changes in chromosome numbers, chromosome rearrangements, gene deletions/amplifications, and conventional point mutations which gradually accumulate during growth under culture conditions [15]. The activation of quiescent transposable elements in chromosomal DNA by the ''stress'' of culturing has also been proposed as a mechanism [99,100]. Whatever the causes, the result is that many cell cultures are highly heterogeneous for any one character. Examples of such heterogeneity include cell-to-cell variation in the level of alkaloids in *Peganum harmala* callus [12], and cultures of Paul's Scarlet Rose (*Rosa* sp.) which contain cells which have either little pigment, general polyphenolics, or else anthocyanins [89]. Given that cultures will be heterogeneous, then it should be possible to isolate sublines which show improved characteristics. This is usually carried out by using fine cell suspensions or protoplast preparations to ensure that the new sublines are of single-cell origin, and thus have defined characteristics. Two basic approaches can be taken—namely, passive screening of cell isolates or active selection for cells with the desired characteristics.

A. Screening

Although initially technically demanding, recent developments [see, e.g., 101] have meant that it is now, in many cases, possible to grow on single cells or protoplasts to form new colonies with reasonable efficiency. In screening procedures large numbers of cells or microcolonies in the process of regrowth are plated out on agar and allowed to grow on to form individual discreet calli. The calli showing the most favorable characteristics are then chosen manually and used to establish large-scale cultures. Recent developments in the field of fluorescence-activated cell sorting [102] also suggest that it should eventually be possible to develop automated systems operating at the single-cell level.

Where one is interested in the level of accumulation of colored or fluorescent products, identification of interesting calli can be done simply by eye. Improving shikonin (a red napthaquinone) production by cultures of *Lithospermum erythrorhizon* has been done in this way [10], and yields have been increased sufficiently (to approximately 20% of cell dry weight [103]) to make shikonin production commercially viable [5]. When the product of interest is colorless, then chemical analysis of the calli is necessary. Radioimmunoassay and enzyme-linked immunosorbent assay

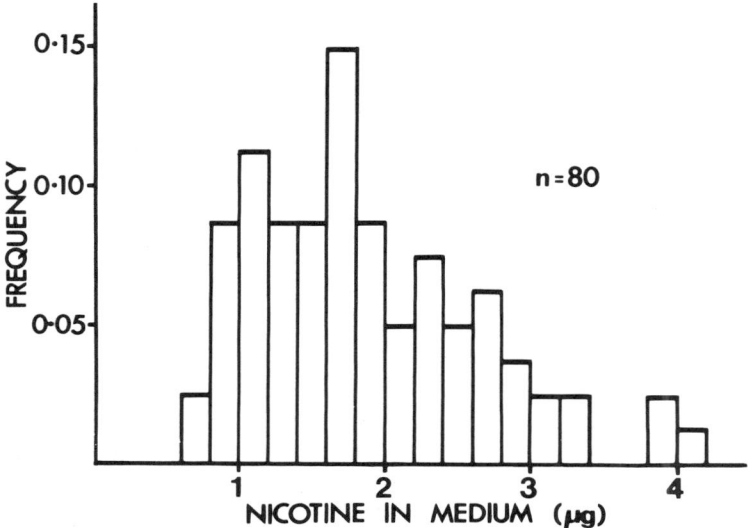

Fig. 2. Variation in the amount of nicotine released to the growth medium by 80 callus-derived "hairy root" lines of *Nicotiana rustica* transformed with *Agrobacterium rhizogenes*. Transformed roots were induced to callus by growth on agar medium containing 0.5 mg/liter benzyladenine and 2 mg/liter napthyleneacetic acid. Following transfer to hormone-free agar, roots regenerated at the surface of the callus. Eighty 1-cm root tips were each transferred to 2 ml growth medium (Gamborg's B5 salts, 2% sucrose, no added hormones) and left to grow on. After 7 days the amount of nicotine in the medium was determined.

(ELISA) [104,105] have proved very valuable here, since they can be automated to cope with the large number of clones generated. Their high sensitivity also means calli can be examined at an early stage, thus producing further savings in time. One slight drawback in comparison to other methods such as high-performance liquid chromatography is that immunological methods tend to be highly specific. With many antibodies, one screening therefore generally detects either a single compound, or a small group of compounds. Potentially interesting clones producing other related compounds (e.g., those that accumulate high levels of metabolic intermediates rather than the final end product) might be missed unless further screenings are done, or an antibody with broader specificity is used.

Figure 2 depicts the sort of variation which is detected in screening experiments. Although this particular example actually relates to variation in the capacity of lines to release products to the medium, the picture is almost identical when the capacity of lines to synthesize products is being examined. Thus, the product content of cell colonies shows a wide variation, with the frequency distribution being skewed towards high values. In some systems, such as solasonine production in *Solanum laciniatum,* about 5% of the colonies can have a content as much as ten times higher than the most common value [106]. It is from such high-producing outliners that large-scale cultures are developed for further work. With luck some of these clones will show stable high production [10,19], but there seems to be a tendency for the behavior of many lines to revert towards lower production [12,107]. This is associated with a redevelopment of cell-to-cell heterogeneity [10,107] and may thus relate to continuing culture-induced variation. Repeated screening and reselection has, however, been shown to stabilize variants in some instances [10], so the problem of instability may not be insurmountable. Since there are indications that high-frequency variation also seems to be characteristic of dispersed cell cultures and less so of organized tissues [97,108], the *Agrobacterium rhizogenes*-transformed root system might also offer a means to stabilizing particular genetic combinations. Cultures could be transformed, and then by appropriate hormone treatments (Hamill, unpublished results) [46] maintained as callus rather than roots. This should enhance spontaneous variation and also allow mutagenesis if required. Roots could then be regenerated directly from this callus, but preferably single cell clones could be isolated following protoplasting, and these clones then allowed to regenerate roots in hormone-free medium in order to return to the organized state. The different stabilized root clones could then be screened in order to identify useful lines. Preliminary work in our laboratory has shown all these steps to be feasible. Much further work is, however, still needed. It is not, for instance, known how much variation will be selected against by the need to regenerate roots. Such effects may be more important for variation in developmental regulation and primary metabolism than for variation in secondary metabolism.

B. Selection

The process of screening requires physical examination of every clone produced, and is thus demanding in terms of labor, space, and time. Many workers have operated on the scale of only approximately 100–200 clones per experiment. By extensive automation it should be possible to increase the number of clones screened to 1000 or more per day [104]. If the improvement being sought is produced only by a rare mutation, even this number may not, however, guarantee success in obtaining a suitable clone. To overcome this problem it would be useful to be able specifically to select out clones with the desired properties. The most obvious approach would be to exert a selection pressure on a cell culture, such that only cells with the required properties are able to divide and grow. A limited number of cells could thus be rapidly singled out from among millions of others. Although immensely successful in the bacterial field, there are several problems in applying this approach to plant cells. There might be problems in growing on the selected cells in the presence of toxic products released from neighboring cells [109], and the growth of cells at low cell density is not always straightforward [110 etc.]. The largest problem is, however, that secondary metabolism is not vital to cell survival, and thus it is not easy to develop suitable selection pressures.

One approach to cell selection which has been applied to secondary metabolism is based on the fact that the initial precursor of most secondary products are key primary metabolites such as amino acids. Many toxic or growth-inhibitory amino acid analogs are known, and these can be used as selective agents. One mode of resistance to amino acid analogs is that the analogs can be ''diluted out'' by overproduction of the correct natural amino acid. Scott et al. [111] have obtained 5-methyltryptophan-resistant lines of *Catharanthus roseus* which overaccumulate tryptophan, owing to altered anthranilate synthase activity which is less subject to feedback inhibition by tryptophan than in typical lines [111]. It has been hoped that overproduction of the primary precursors might be associated with increased secondary product synthesis. In practice the results are, however, rarely encouraging [111 etc.],,although very slight increases might occur. The highest level of alkaloid in cultures of *C. roseus* examined by Deus-Neumann and Zenk [107] was thus seen in a 5-methyltryptophan-resistant line, but levels were only 35% greater than in some nonresistant lines. The general applicability of the concept of overproduction is in fact open to question, since supply of primary precursors need not be rate-limiting to secondary product formation, and also one way to increase precursor levels may indeed be to reduce the flux into secondary products! A more productive approach to take seems to be selection for metabolic detoxification. If the substrate analog fed can be metabolized by the enzymes of the secondary product pathway of interest, and if the products are less toxic than the initial analog, then it should be possible to select for cells showing enhanced activity of the initial enzymes of the pathway. This is because these cells will be more efficient detoxifiers of the selective agent, and will thus be resistant to higher concentration than other cells. Since early enzymes are likely to be rate-limiting to a metabolic pathway [112], this type of selection should stand a good chance of producing lines with enhanced secondary product synthesis. Sasse et al. [113] have obtained 4-methyltryptophan-resistant lines of *C. roseus* with enhanced tryptophan decarboxylase activity by this method, and Palmer and Widholm [114] have obtained a p-fluorophenylalanine-resistant line of *Nicotiana tabacum* which among other alterations possesses enhanced phenylalanine ammonia-lyase activity [115]. These *Nicotiana* cells synthesize large amounts of fluorinated phenolic compounds when fed p-fluorophenylalanine [116] and accumulate tenfold higher levels of phenolics than unselected lines even in the absence of the section pressure [117]. Similar increases in phenolic production have also been observed in p-fluorophenylalanine-resistant lines of *Acer pseudoplatanus* [118].

As mentioned earlier, one of the problems of high productivity lines is that they are often unstable and tend to revert to lower productivity. Lines obtained by selection are not immune to this problem, and indeed it is possible to obtain many ''adapted'' lines as well as true variants. These lines grow during the initial selection step, but lose the selected phenotype as soon as the selection pressure is removed [119 etc.]. The advantage of selection procedures in the search for stability is that lines can be readily reselected at intervals to help either stabilize the system or else simply remove undesirable revertants at each stage. For example, Schiel et al. [120] found that a p-fluorophenylalanine-resistant line of *N. tabacum* which had been grown in the absence of selection pressure for over a year had begun to lose its enhanced capacity to synthesize cinnamoylputrescines. The high-level productivity could, however, be restored by growing the cells for two cycles in the presence of p-fluorophenylalanine.

C. Genetic Manipulation

Another way to improve secondary product production by cultured cells would be to engineer cell lines genetically by the introduction of foreign DNA. Techniques for this, based on *Agrobacterium* vectors [121] or direct DNA uptake [122], are now well established. At present there are, however, still major limitations, particularly with the more biochemical aspects. Thus, the factors governing the flux through secondary product pathways are rarely understood at either the molecular or biochemical level (see section II.H for some exceptions), and since there can be significant turnover of some secondary products [123], there is no guarantee that increasing synthetic rates will always increase accumulation. In addition, so little biochemical work has been done on the enzymology of many important secondary product pathways that the identification and isolation of genes worth transferring is far from straightforward. Progress, however, continues to be made, particularly in the field of alkaloids [124], and in a few years' time genetic engineering should become a useful tool in the manipulation of secondary product synthesis by plant cells in culture.

IV. EXPLOITATION SYSTEMS

To be a viable alternative to conventional agriculture in the production of secondary products, plant cell cultures must not only produce the relevant products, but must also be capable of being exploited in an efficient manner. There are two aspects to this—namely, the growth of large quantities of cells, and the recovery of the desired products.

A. Fermenter Systems for Growing Biomass
1. Free cell reactors. There is now considerable expertise in growing freely suspended cells in large-scale fermenters [see, e.g., 125,126]. Since plant cells are relatively shear sensitive [127], airlift systems [128,129] with their gentle aeration and mixing are often preferred, although impellor-stirred systems can also be used sucessfully in a number of cases. Culture vessels with capacities as large as 20,000 liters have been investigated [126], and cell densities as high as 30 g dry weight per liter have sometimes been obtained [127].

For cells which are producing high-value products, free cell fermenters offer an effective exploitation system. Cells are grown as a batch either directly in production medium, or else initially in growth medium followed by a change to production medium once a high cell mass has been obtained (cf. section II.A.). After a suitable period—which will depend on whether product synthesis is growth related, or occurs primarily once growth slows— the cells and medium are then harvested and product extracted. The Japanese process for shikonin production works on this basis [5].

2. Immobilized cell reactors. Free cell reactors have a number of technical features which may sometimes prove to be inconvenient. For instance, they generally have to be run in a discontinuous batch mode, rather than in a continuous mode. This is because plant cells grow very slowly, with generation times in the range of 15 hours [126] to 10 days [130] or even longer. This means that continuous-flow systems are generally impractical since the dilution rates that can be achieved without ultimate washout of the cells are correspondingly low. One solution to this problem is the use of immobilized, rather than free, cells in the fermenter. The cell aggregates have sufficient size and density to be retained in the fermenter under high dilution rates, and provided cells release some of their product into the medium (see IV.B. below) a continuous harvesting of product from the effluent is possible. As mentioned earlier, the use of immobilized cells may also offer certain biochemical advantages in that their productivity may be greater than that of free cells. In addition, it appears that immobilization in many cases stabilizes cells so that their biosynthetic capacity survives longer under nongrowing conditions than in equivalent free cells [33]. This allows more effective utilization of the generated biomass. While free cell systems are in many cases perfectly adequate for exploitation of cells, it is clearly apparent that immobilized cell systems have a number of additional advantages for the commercial exploitation of many systems. There is currently much interest in immobilized cells [e.g., 32,34,35,41]; and it is hoped that with time the technology will allow the economic production of compounds whose present price is as little as $50 per kilogram for the agriculturally produced or synthetic product [38].

Many of the laboratory-scale immobilized cell reactors currently in operation are based upon airlift fermentors containing cells immobilized on reticulated foams [42,131]. Although each foam particle can be nearly totally filled with viable cells [39,40], there is a limit to the cell density which can be achieved, since the energy needed to suspend the particles and aerate the system increases rapidly once particles occupy more than about one-third of the fermenter volume [42,131]. The limiting cell density is, however, close to that commonly obtained from free cell suspensions, so this is not likely to be a serious limitation. If necessary, a column of foam particles can be used, rather than a circulating bed system.

The other major systems operating on a laboratory scale are based on cells immobilized in gels, e.g., alginate beads [35] or polyacrylamide sheets [132]. These systems suffer from the problem that biomass has to be first generated before the cells are subsequently immobilized. Also, the gel matrices tend to be somewhat compressible, so that they may conceivably collapse under loading in large columns. It thus seems likely that these systems might prove more difficult to scale-up than those based on the in situ growth of cells inside reticulated foams.

3. Transformed root cultures. *Agrobacterium rhizogenes*-transformed roots are in many cases very robust, and can be grown directly in fermenters. In many respects their culture properties show a mixture of characteristics of both free and immobilized cells. They can be grown as a batch to fill nearly the entire fermenter volume [47], and then by connecting up a flow of medium, the fermenter can be coverted to a continuous mode [47]. Cell densities as high as 40 g dry weight per liter have been obtained in our laboratory (Hilton et al., unpublished observations).

B. Harvesting of Products

1. Products spontaneously released to medium. Many secondary products are spontaneously released by plant cells grown in culture, as illustrated in Table IV. This allows the cells to be exploited in long-term continuous-flow-through systems, with product being harvested from the medium.

TABLE IV. Examples of Secondary Products Which Can in Some Cases be Spontaneously Released by Plant Cells in Culture*

Alkaloids
Benzylisoquinolines (e.g., berberine, protopine)
Indoles (e.g., ajmalicine)
Pyridines (e.g., nicotine)
Quinolines (e.g., quinine)
Quinolizidines (e.g., lupanine)
Anthraquinones (e.g., purpurin-1-methyl ether)
Capsaicin
L-DOPA
Opines (e.g., nopaline)
Phenolics
Some simple phenolics
Coumarins
Terpenoids
Monoterpenoids (e.g., terpinolene)
Sesquiterpenoids (e.g., paniculide B)

*See references [21, 25, 33, 36, 45, 50, 52, 133, 136, 139, 141, 144].

Obviously if this approach is used it will be desirable to maximize product release, since this now becomes the main determinant of daily productivity. A number of strategies are available for doing this. First, it has been shown that the extent of release of a given compound can vary between different species [133], or even between different lines of the same species [134] (see also section III.A.). The choice of the right starting material—not only in terms of net productivity, but also in terms of the extent of product release— is thus an important factor in establishing a commercially viable process. Once a suitable line has been identified it may be possible to further improve product release by manipulating environmental conditions. Passive diffusion is believed to be important in the transport of some alkaloids [135,136], and possibly other secondary products. Since neutral molecules are generally much more membrane permeant than ions, if the product of interest has an ionizable group (as do most alkaloids), then the rate of diffusion and the final distribution of product between cells and medium should be sensitive to intracellular and extracellular pH. In addition to simple diffusion there is now growing evidence that specific carriers may also be involved in the transport of a variety of compounds including some alkaloids [137,138], and probably also capsaicin [36,139]. If carrier proteins are involved in product secretion, it should be possible to manipulate the activity of these carriers and so influence product release. At present such alterations will have to be obtained on a purely empirical basis, except in rare cases where it might be possible directly to select cell lines showing efficient release (see ref. 140 for an example relating to a primary metabolite) or where enough is known about the molecular biology for genetic engineering to become feasible, as is the case for opine secretion in *Agrobacterium tumefaciens*-transformed cells [141].

In addition to the manipulations described above, it should be kept in mind that process design can also influence product release to the medium. A flow-through system in which cells are continually exposed to fresh medium increases product release by ensuring that the system is always maintained away from equilibrium. Increased flow rates are associated with increased product recovery [47], although in some systems the need for cells to condition their medium [142] may restrict the flow rates which can be used. An alternative approach to stimulating product release is to have an absorbent present in the medium. This binds product, and so keeps the concentration of free product in the medium very low. This in turn is beneficial to product release. The use of a specific absorbent has, in fact, much to recommend it. Not only will it enhance release of product from cells, but there may also be other biological advantages. The product might be stabilized against biodegradation, which can sometimes be a problem when products occur in the medium [11,143]. Also, the continual removal of product may lead to stimulation of total synthesis, as seen for anthraquinone production in *Cinchona ledgeriana* by Robins and Rhodes [144]. This might conceivably relate to the relief of feedback controls which operate in unmanipulated cultures. Finally, it should be noted that specific absorbents effectively concentrate up compounds out of solution, and so should also facilitate the final extraction and purification of products.

2. Products not normally released to medium. While many products are spontaneously released by cells, there are also a large number which are not released, but which are instead sequestered within the cells, usually within the vacuole [145,146]. If the cells are to be exploited by a batch culture process this is not a problem, since cells can be harvested at the end of the run, and the products then extracted. In some situations it might, however, be desirable to harvest the products without having to destroy the cells. In these circumstances the goal would be to permeablize the cells so that products diffuse into the medium without vital metabolites also being lost, or else the cells retain an ability to restore their integrity once the permeabilizing agent is withdrawn. A wide variety of treatments are known which permeabilize cells and cause release of products, and these have been reviewed by Felix [147]. Many treatments, such as most organic solvents, are very harsh and totally disrupt membranes. There are, however, reports of successful permeabilization with maintenance of viability. Fuller and Bartlett [30] found that when salicylic acid is fed to cells of *Populus alba* it is glucosylated, but only in the presence of the detergent cetyl-trimethyl ammonium bromide did the product appear in the medium. Brodelius and Nilsson [148] were able to permeabilize cells of *Catharanthus roseus* with 5% (v/v) dimethylsulphoxide (DMSO) for 30 min, so that the cells released almost all of their internal alkaloids. When the DMSO was removed the cells resumed alkaloid synthesis, and after a few days a further brief treatment with DMSO could be used to release the stored alkaloid [148]. Recently Tanaka et al. [149] have also reported that high-ionic-strength media can permeabilize *C. roseus* cells with maintenance of viability.

Despite the reports of successful permeabilization procedures, the generality of these are still open to question. Although DMSO reversibly permeabilized *C. roseus* cells [148], it has been shown to produce irreversible cell damage in *Cinchona ledgeriana* [150] and *Berberis* [151]. Since plant vacuoles have a relatively low internal pH compared to the cytoplasm [152,153], and often have a high content of phenolics [154,155], degradative enzymes [156,157], and other potential toxins, it is perhaps not surprising that permeabilization of the tonoplast in order to release vacuolar stored products can lead to cell damage. Some of the reports of successful cell permeabilization with maintenance of viability may thus have been interpreted too simply. The fact that, whatever the explanation, there do seem to be some workable systems, shows that cell permeabilization techniques can, however, be of potential importance in the development of commercial systems. Further improvements based upon continuing research are also to be expected [139].

V. CONCLUSIONS

As well as providing food and bulk materials, plants are a valuable source of high-cost, low-volume compounds used in the pharmaceutical, food, and fragrance industries. Because of their complex chemical structures with many chiral centers, these compounds are often difficult to prepare synthetically, and conventional agriculture is still the main source of these products. The use of plant cell cultures to supplement or replace this supply is of great commercial interest. Two basic problems exist in the development of such cell culture systems—namely, how to obtain cells with an economically viable yield of the desired product, and how best to exploit these cells. In recent years significant developments have been made in both aspects. Finding effective ways of enhancing yield is perhaps the key to success, since this is crucial to whether a process will ever be economic. In a number of systems it is clear that very substantial advances have been made, and cultures may now often synthesize higher levels of secondary products than the parent plant. Many of these advances have been obtained empirically, but our understanding of plant cell systems continues to grow, and it should soon be possible to develop a rational approach to manipulation of cell cultures. Although only a limited number of processes, such as that for shikonin, are currently at a commercial stage, there is optimism that the production of secondary products from plant cell cultures will soon come of age.

REFERENCES

1. Treimer JF, Zenk MH: Strictosidine synthase from cell cultures of Apocynaceae plants. FEBS Lett 97:159, 1979.
2. Bell EA: The possible significance of secondary compounds in plants. In Bell EA, Charlwood BV (eds): "Encyclopedia of Plant Physiology. Volume 8—Secondary Plant Products." Berlin: Springer-Verlag, 1980, p 11.
3. Bell EA: The Physiological role(s) of secondary (natural) products. In Conn EE (ed): "The Biochemistry of Plants—A Comprehensive Treatise. Volume 7—Secondary Plant Products." New York: Academic Press, 1981, p 1.
4. Balandrin MF, Klocke JA, Wurtele ES, Bollinger WH: Natural plant chemicals—sources of industrial and medicinal materials. Science 228:1154, 1985.
5. Curtin ME: Harvesting profitable products from plant tissue culture. Bio/technology 1:649, 1983.
6. Murashige T, Skoog F: A revised medium for rapid growth and bioassays with tobacco tissue cultures. Physiol Plant 15:473, 1962.
7. Gamborg OL, Miller RA, Ojima K: Nutrient requirements of suspension cultures of soybean root cells. Exp Cell Res 50:151, 1968.
8. Linsmaier EM, Skoog F: Organic growth factor requirements of tobacco tissue cultures. Physiol Plant 18:100, 1965.
9. Schenk RU, Hildebrandt AC: Medium and techniques for induction and growth of monocotyledonous and dicotyledonous plant cell cultures. Can J Bot 50:199, 1972.
10. Tabata M, Ogino T, Yoshioka K, Yoshikawa N, Hiraoka N: Selection of cell lines with higher yield of secondary products. In Thorpe TA (ed): "Frontiers of Plant Tissue Culture 1978." University of Calgary: International Association for Plant Tissue Culture, 1978, p 213.
11. Robins RJ, Furze JM, Rhodes MJC: α-Acid degradation by suspension culture cells of *Humulus lupulus*. Phytochemistry 24:709, 1985.
12. Sasse F, Heckenberg U, Berlin J: Accumulation of β-carboline alkaloids and serotonin by cell cultures of *Peganum harmala*. I. Correlation between plants and cell cultures and influence of medium constituents. Plant Physiol 69:400, 1982.
13. Dhoot GK, Henshaw GG: Organization and alkaloid production in tissue cultures of *Hyoscyamus niger*. Ann Bot 41:943, 1977.
14. Bayliss MW: Chromosomal variation in plant tissues in culture. In Vasil IK (ed): "Perspectives in Plant Cell and Tissue Culture, International Review of Cytology Supplement 11A." New York: Academic Press, 1980, p 113.

15. Evans DA, Sharp WR: Applications of somaclonal variation. Bio/technology 4:528, 1986.
16. Andersen RA, Kemp TR, Vaughn TH: Coniferyl alcohol, sinapyl alcohol and scopoletin in tobacco callus tissue during growth of a subculture. Physiol Plant 53:89, 1981.
17. Misawa M, Hayashi M, Takayama S: Accumulation of antineoplastic agents by plant tissue cultures. In Neumann K-H, Barz W, Reinhard E (eds): "Primary and Secondary Metabolism of Plant Cell Cultures." Berlin: Springer-Verlag, 1985, p 235.
18. Matsumoto T, Nishida K, Noguchi M, Tamaki E: Some factors affecting the anthocyanin formation by *Populus* cells in suspension culture. Agr Biol Chem 37:561, 1973.
19. Zenk MH, El-Shagi H, Arens H, Stöckigt J, Weiler EW, Deus B: Formation of the indole alkaloids serpentine and ajmalicine in cell suspension cultures of *Catharanthus roseus*. In Barz W, Reinhard E, Zenk MH (eds): "Plant Tissue Culture and Its Biotechnological Application." Berlin: Springer-Verlag, 1977, p 27.
20. Staba EJ: Secondary metabolism and biotransformation. In Staba EJ (ed): "Plant Tissue Culture as a Source of Biochemicals." Boca Raton. FL: CRC Press, 1980, p 59.
21. Böhm H: Regulation of alkaloid production in plant cell cultures. In Thorpe TA (ed): "Frontiers of Plant Tissue Culture 1978." University of Calgary: International Association for Plant Tissue Culture, 1978, p 201.
22. Fahn W, Gundlach H, Deus-Neumann B, Stöckigt J: Late enzymes of vindoline biosynthesis. Acetyl-CoA: 17-O-deacetylvindoline 17-O-acetyl-transferase. Plant Cell Rep 4:333, 1985.
23. Kurz WGW, Chatson KB, Constabel F, Kutney JP, Choi LSL, Kolodziejczyk P, Sleigh SK, Stuart KL, Worth BR: Alkaloid production in *Catharanthus roseus* cell cultures—initial studies on cell lines and their alkaloid content. Phytochemistry 19:2583, 1980.
24. Böhm H: The formation of secondary metabolites in plant tissue and cell cultures. In Vasil IK (ed): "Perspectives in Plant Cell and Tissue Culture, International Review of Cytology Supplement 11B." New York: Academic Press, 1980, p 183.
25. Butcher DN, Connolly JD: An investigation of factors which influence the production of abnormal terpenoids by callus cultures of *Andrographis paniculata* Nees. J Exp Bot 22:314, 1971.
26. Tabata M, Yamamoto H, Hiraoka N, Konoshima M: Organisation and alkaloid production in tissue cultures of *Scopolia parviflora*. Phytochemistry 11:949, 1972.
27. Koul S, Ahuja A, Grewal S: Growth and alkaloid production in suspension cultures of *Hyoscyamus muticus* as influenced by various cultural parameters. Planta Med 47:11, 1983.
28. Hagimori M, Matsumoto T, Kisaki T: Studies on the production of *Digitalis* cardenolides by plant tissue culture. I. Determination of digitoxin and digoxin contents in first and second passage calli and organ redifferentiating calli of several *Digitalis* species by radioimmunoassay. Plant Cell Physiol 21:1391, 1980.
29. Dhir SK, Shekhawat NS, Purohit SD, Arya HC: Development of laticifer cells in callus cultures of *Calotropis procera* (Ait.) R.Br. Plant Cell Rep 3:206, 1984.
30. Fuller KW, Barlett DJ: The chemosynthetic potential of plants and its realisation by immobilized systems. In Fuller KW, Gallon JR (eds): "Plant Products and the New Technology," Ann Proc Phytochem Soc Eur, Vol 26. Oxford: Oxford University Press, 1985, p 229.
31. Hay CA, Anderson LA, Roberts MF, Phillipson JD: In vitro cultures of *Cinchona* species. Precursor feeding of *C. ledgeriana* root organ suspension cultures with L-tryptophan. Plant Cell Rep 5:1, 1986.
32. Lindsey K, Yeoman MM: Novel experimental systems for studying the production of

secondary metabolites by plant tissue cultures. In Mantell SH, Smith H (eds): "Plant Biotechnology," Soc Exp Bot Seminar Series, 18. Cambridge: Cambridge University Press, 1983, p 39.

33. Rhodes MJC: Immobilized plant cell cultures. In Wiseman A (ed): "Topics in Enzyme and Fermentation Biotechnology"—Vol. 10. Chichester: Ellis Horwood Ltd., 1985, p 51.
34. Rosevear A, Lambe CA: Immobilized plant cells. Adv Biochem Eng Biotechnol 31:37, 1985.
35. Brodelius P, Mosbach K: Immobilized plant cells. Adv Appl Microbiol 28:1, 1982.
36. Lindsey K, Yeoman MM, Black GM, Mavituna F: A novel method for immobilization and culture of plant cells. FEBS Lett 155:143, 1983.
37. Rhodes MJC, Robins RJ, Turner RJ, Smith JI: Mucilaginous film production by plant cells immobilized in a polyurethane or nylon matrix. Can J Bot 63:2357, 1985.
38. Rhodes MJC, Kirsop BH: Plant cell cultures as sources of valuable secondary products. Biologist 29:134, 1982.
39. Parr AJ, Robins RJ, Rhodes MJC: Apparent free space and cell volume estimation—A non-destructive method for assessing the growth and membrane integrity/viability of immobilized plant cells. Plant Cell Rep 3:161, 1984.
40. Mavituna F, Park JM: Growth of immobilized plant cells in reticulate polyurethane foam matrices. Biotechnol Lett 7:637, 1985.
41. Rhodes MJC, Smith JI, Robins RJ: Factors affecting the immobilization of plant cells on reticulated polyurethane foam particles. Appl Microbiol Biotechnol 26:28, 1987.
42. Mavituna F, Park JM, Williams PD, Wilkinson AK: Characteristics of immobilized plant cell reactors. In Webb C, Mavituna F (eds): "Process Possibilities for Plant and Animal Cell Cultures," I Chem E Symposium, UMIST 1986. Chichester: Ellis Horwood Ltd., 1987, p 92.
43. Lindsey K, Yeomann MM: Immobilized plant cell culture systems. In Neumann K-H, Barz W, Reinhard E (eds.): "Primary and Secondary Metabolism of Plant Cell Cultures." Berlin: Springer-Verlag, 1985, p 304.
44. Chilton M-D, Tepfer DA, Petit A, David C, Casse-Delbart F, Tempé J: *Agrobacterium rhizogenes* inserts T-DNA into the genomes of the host plant root cells. Nature 295:432, 1982.
45. Hamill JD, Parr AJ, Robins RJ, Rhodes MJC: Secondary product formation by cultures of *Beta vulgaris* and *Nicotiana rustica* transformed with *Agrobacterium rhizogenes*. Plant Cell Rep 5:111, 1986.
46. Flores HE, Filner P: Metabolic relationships of putrescine, GABA and alkaloids in cell and root cultures of Solanaceae. In Neumann K-H, Barz W, Reinhard E (eds): "Primary and Secondary Metabolism of Plant Cell Cultures." Berlin: Springer-Verlag, 1985, p 174.
47. Rhodes MJC, Hilton M, Parr AJ, Hamill JD, Robins RJ: Nicotine production by "hairy root" cultures of *Nicotiana rustica*—fermentation and product recovery. Biotechnol Lett 8:415, 1986.
48. Ooms G, Hooykaas PJJ, Moolenaar G, Schilperoort RA: Crown gall plant tumors of abnormal morphology, induced by *Agrobacterium tumefaciens* carrying mutated octopine Ti plasmids; analysis of T-DNA functions. Gene 14:33, 1981.
49. Nestor EW, Gordon MP, Amasino RM, Yanofsky MF: Crown gall—a molecular and physiological analysis. Annu Rev Plant Physiol 35:387, 1984.
50. Brain KR: Accumulation of L-DOPA in cultures from *Mucuna pruriens*. Plant Sci Lett 7:157, 1976.

51. Lindsey K, Yeomann MM: The relationship between growth rate, differentiation and alkaloid accumulation in cell cultures. J Exp Bot 34:1055, 1983.
52. Robins RJ, Payne J, Rhodes MJC: Cell suspension cultures of *Cinchona ledgeriana*. I. Growth and quinoline alkaloid production. Planta Med 220, 1986.
53. Zenk MH, El-Shagi H, Schulte U: Anthraquinone production by cell suspension cultures of *Morinda citrifolia*. Planta Med [Suppl]:79, 1975.
54. Kurz WGW, Constabel F: Plant cell cultures, a potential source of pharmaceuticals. Adv Appl Microbiol 25:209, 1979.
55. Rhodes MJC, Payne J, Robins RJ: Cell suspension cultures of *Cinchona ledgeriana*. II. The effect of a range of auxins and cytokinins on the production of quinoline alkaloids. Planta Med 226, 1986.
56. Kirsop BA, Rhodes MJC, Robins RJ: Improvements in the production of metabolites by cell culture. Br Patents 129259 and 129260; 1986.
57. van Geyt JPC, Jacobs M: Suspension culture of sugarbeet (*Beta vulgaris* L.). Induction and habituation of dedifferentiated and self-regenerating cell lines. Plant Cell Rep 4:66, 1985.
58. Payne J, Rhodes MJC, Robins RJ: Quinoline alkaloid production by transformed cultures of *Cinchona ledgeriana*. Planta Med 53:367, 1987.
59. Norton RA, Finlayson AJ, Towers GHN: Thiophene production by crown galls and callus tissues of *Tagetes patula*. Phytochemistry 24:719, 1985.
60. Mérillon JM, Doireau P, Guillot A, Chénieux JC, Rideau M: Indole alkaloid accumulation and tryptophan decarboxylase activity in *Catharanthus roseus* cells cultured in three different media. Plant Cell Rep 5:23, 1986.
61. Dougall DK: Nutrition and metabolism. In Staba EJ (ed): "Plant Tissue Culture as a Source of Biochemicals." Boca Raton, FL: CRC Press, 1980, p 21.
62. Schulte U, El-Shagi H, Zenk MH: Optimization of 19 Rubiaceae species in cell culture for the production of anthraquinones. Plant Cell Rep 3:51, 1984.
63. Ritchie G: From discovery to commercial reality, some aspects of fermentation product development. Chem Ind 403, 1985.
64. Schiel O, Jarchow-Redecker K, Piehl G-W, Lehmann J, Berlin J: Increased formation of cinnamoyl putrescines by fedbatch fermentation of cell suspension cultures of *Nicotiana tabacum*. Plant Cell Rep 3:18, 1984.
65. Knobloch K-H, Beutnagel G, Berlin J: Influence of accumulated phosphate on culture growth and formation of cinnamoyl putrescines in medium-induced cell suspension cultures of *Nicotiana tabacum*. Planta 153:582, 1981.
66. Knobloch K-H, Belin J: Influence of medium composition on the formation of secondary compounds in cell suspension cultures of *Catharanthus roseus* (L.) G Don. Z Naturforsch 35C:551, 1980.
67. Dear J, Aronoff S: Relative kinetics of chlorogenic and caffeic acids during the onset of boron deficiency in sunflower. Plant Physiol 40:458, 1965.
68. Dixon RA: Plant tissue culture methods in the study of phytoalexin induction. In Ingram DS, Helgeson JP (eds): "Tissue Culture Methods for Plant Pathologists." Oxford: Blackwell's, 1980, p 185.
69. DiCosmo F, Misawa M: Eliciting secondary metabolism in plant cell cultures. Trends Biotechnol 3:318, 1985.
70. Kuhn DN, Chappell J, Boudet A, Hahlbrock K: Induction of phenylalanine ammonia-lyase and 4-coumarate: CoA ligase mRNAs in cultured plant cells by UV light or fungal elicitor. Proc Natl Acad Sci USA 81:1102, 1984.
71. Schmelzer E, Somssich I, Hahlbroch K: Coordinated changes in transcription and

translation rates of phenylalanine ammonia-lyase and 4-coumarate: CoA ligase mRNAs in elicitor-treated *Petrolselinum crispum* cells. Plant Cell Rep 4:293, 1985.

72. Wijnsma R, Go JTKA, van Weerden IN, Harkes PAA, Verpoorte R, Svendsen AB: Anthraquinones as phytoalexins in cell and tissue cultures of *Cinchona* spec. Plant Cell Rep 4:241, 1985.

73. Rokem JS, Schwarzberg J, Goldberg I: Autoclaved fungal mycelia increase diosgenin production in cell suspension cultures of *Dioscorea deltoidea*. Plant Cell Rep 3:159, 1984.

74. Tietjen KG, Hinkler D, Matern U: Differential response of cultured parsley cells to elicitors from two non-pathogenic strains of fungi. Eur J Biochem 131:401, 1983.

75. Lindsey K, Yeoman MM: The synthetic potential of immobilized cells of *Capsicum frutescens* Mill cv. annuum. Planta 162:495, 1984.

76. Deus B, Zenk MH: Exploitation of plant cells for the production of natural compounds. Biotechnol Bioeng 24:1965, 1982.

77. Robins RJ, Hanley AB, Richards SR, Fenwick RG, Rhodes MJC: Uncharacteristic alkaloid synthesis by suspension cultures of *Cinchona pubescens* fed with L-tryptophan. Plant Cell Tissue Org Cult 9:49, 1987.

78. Reinhard E, Alfermann AW: Biotransformation by plant cell cultures. Adv Biochem Eng Biotechnol 16:49, 1980.

79. Stohs SJ: Metabolism of steroids in plant tissue cultures. Adv Biochem Eng Biotechnol 16:85, 1980.

80. Martin SM: Environmental factors. B. Temperature, aeration and pH. In Staba EJ (ed): "Plant Tissue Culture as a Source of Biochemicals." Boca Raton, FL: CRC Press, 1980, p 143.

81. Smart NJ, Fowler MW: Effect of aeration on large-scale cultures of plant cells. Biotechnol Lett 3:171, 1981.

82. Scragg AH, Morris P, Allan EJ, Bond P, Hegarty P, Smart NJ, Fowler MW: The effect of scale-up on plant-cell culture performance. In Webb C, Mavituna F (eds): "Process Possibilities for Plant and Animal Cell Culture," I Chem E Symposium, UMIST 1986. Chichester: Ellis Horwood Ltd., 1987, p 78.

83. Breuling M, Alfermann AW, Reinhard E: Cultivation of cell cultures of *Berberis wilsonae* in 20-1 airlift bioreactors. Plant Cell Rep 4:220, 1985.

84. Hazell LP: Data quoted by Mantell SH, Smith H—Cultural factors that influence secondary metabolite accumulations in plant cell and tissue culture. Soc Exp Bot Seminar Series 18:75, 1983. [see reference 95.]

85. Nettleship L, Slaytor M: Adaption of *Peganum harmala* callus to alkaloid production. J Exp Bot 25:1114, 1974.

86. Seibert M, Kadkade PG: Environmental factors. A. Light. In Staba EJ (ed): "Plant Tissue Culture as a Source of Biochemicals." Boca Raton, FL: CRC Press, 1980, p 123.

87. Hahlbrock K, Knoblock K-H, Kreuzaler F, Potts JRM, Wellmann E: Coordinated induction and subsequent activity changes of two groups of metabolically interrelated enzymes. Light induced synthesis of flavonoid glycosides in cell suspension cultures of *Petroselinum hortense*. Eur J Biochem 61:199, 1976.

88. Gregor H-D, Reinert J: Induktion der phenylalanin-ammonium-lyase in gewebekulturen von *Haplopappus gracilis*. Protoplasma 74:307, 1972.

89. Davis ME: Polyphenol synthesis in cell suspension cultures of Paul's Scarlet Rose. Planta 104:50, 1972.

90. Roller U: Selection of plants and plant tissue cultures of *Catharanthus roseus* with high content of serpentine and ajmalicine. In Alfermann AW, Reinhard E (eds): "Production

of Natural Compounds by Cell Culture Methods." Munich: Gesellschaft fur Strahlen und Umweltforschung m.b.H., 1978, p 95.

91. Corduan G, Reinhard E: Synthesis of volatile oils in tissue cultures of *Ruta graveolens*. Phytochemistry 11:917, 1972,

92. Nagel M, Reinhard E: Das atherische öl der calluskulteren von *Ruta graveolens*. II. Physiologie zur bildung des atherischen öles. Planta Med 27:264, 1975.

93. Tabata M, Mizukami H, Hiraoka N, Konoshima M: Pigment formation in callus cultures of *Lithospermum erythrorhizon*. Phytochemistry 13:927, 1974.

94. Ohta S, Yatazawa M: Effect of light on nicotine production in tobacco tissue culture. Agric Biol Chem 42:873, 1978.

95. Mantell SH, Smith H: Cultural factors that influence secondary metabolite accumulations in plant cell and tissue cultures. In Mantell SH, Smith H (eds): "Plant Biotechnology," Soc Exp Bot Seminar Series, 18. Cambridge: Cambridge University Press, 1983, p 75.

96. Skinner S, Walton NJ, Robins RJ, Rhodes MJC: Tryptophan decarboxylase, strictosidine synthase and alkaloid production by *Cinchona ledgeriana* suspension cultures. Phytochemistry 26:721, 1987.

97. Reisch B: Genetic variability in regenerated plants. In Evans DA, Sharp WR, Ammirato PV, Yamada Y (eds): "Handbook of Plant Cell Culture; Volume 1, Techniques for Propagation and Breeding." New York: Macmillan Publishing Co., 1983, p 748.

98. Kinnersley AM, Dougall DK: Correlation between the nicotine content of tobacco plants and callus cultures. Planta 149:205, 1980.

99. Freeling M: Plant transposable elements and insertion sequences. Annu Rev Plant Physiol 35:277, 1984.

100. Groose RW, Bingham ET: An unstable anthocyanin mutation recovered from tissue culture of alfalfa (*Medicago sativa*) 1. High frequency of reversion upon reculture. Plant Cell Rep 5:104, 1986.

101. Shillito RD, Paszkowski J, Potrykus I: Agarose plating and a bead type culture technique enable and stimulate development of protoplast-derived colonies in a number of plant species. Plant Cell Rep 2:244, 1983.

102. Brown S, Renaudin J-P, Prévot C, Guern J: Flow cytometry and sorting of plant protoplasts—technical problems and physiological results from a study of pH and alkaloids in *Catharanthus roseus*. Physiol Vég 22:541, 1984.

103. Fujita Y, Takahasi S, Yamada Y: Selection of cell lines with high productivity of shikonin derivatives through protoplast of *Lithospermum erythrorhizon*. In "Third European Congress on Biotechnology." Weinheim: Verlag Chemie, 1984, Vol 1, p 161.

104. Weiler EW: Radioimmuno-screening methods for secondary plant products. In Barz W, Reinhard E, Zenk MH (eds): "Plant Tissue Culture and Its Biotechnological Application." Berlin: Springer-Verlag, 1977, p 266.

105. Robins RJ: The measurement of low-molecular weight, nonimmunogenic compounds by immunoassay. In Linskens H-F, Jackson JF (eds): "Modern Methods of Plant Analysis, New Series." Berlin: Springer-Verlag, 1986, Vol 4, p 86.

106. Zenk MH: The impact of plant tissue culture on industry. In Thorpe TA (ed): "Frontiers of Plant Tissue Culture, 1978." University of Calgary: International Association for Plant Tissue Culture, 1978, p 1.

107. Deus-Neumann B, Zenk MH: Instability of indole alkaloid production in *Catharanthus roseus* cell suspension cultures. Planta Med 50:427, 1984.

108. Banerjee-Chattopadhyay S, Schwemmin AM, Schwemmin DJ: A study of karyotypes and their alterations in cultured and *Agrobacterium* transformed roots of *Lycopersicon peruvianum* Mill. Theor Appl Genet 71:258, 1985.

109. Flick CE: Isolation of mutants from cell culture. In Evans DA, Sharp WR, Ammirato PV, Yamada Y (eds): "Handbook of Plant Cell Culture; Volume 1, Techniques for Propagation and Breeding." New York: Macmillan Publishing Co., 1983, p 393.
110. Bellincampi D, Baduri N, Morpurgo G: High plating efficiency with plant cell cultures. Plant Cell Rep 4:155, 1985.
111. Scott AI, Mizukami H, Lee S-L: Characterization of a 5-methyltryptophan resistant strain of *Catharanthus roseus* cultured cells. Phytochemistry 18:795, 1979.
112. Conn EE, Stumpf PK: "Outlines of Biochemistry," Third edition. New York: John Wiley and Sons, 1972, pp 465–486.
113. Sasse F, Buckholz M, Berlin J: Selection of cell lines of *Catharanthus roseus* with increased tryptophan decarboxylase activity. Z Naturforsch 38C:916, 1983.
114. Palmer JE, Widholm J: Characterization of carrot and tobacco cell cultures resistant to p-fluorophenylalanine. Plant Physiol 56:233, 1975.
115. Berlin J, Widholm JM: Corelation between phenylalanine ammonia lyase activity and phenolic biosynthesis in p-fluorophenylalanine-sensitive and -resistant tobacco and carrot tissue cultures. Plant Physiol 59:550, 1977.
116. Berlin J, Witte L, Hammer J, Kukoschke KG, Zimmer A, Pape D: Metabolism of p-fluorophenylalanine in p-fluorophenylalanine sensitive and resistant tobacco cell cultures. Planta 155:244, 1982.
117. Berlin J, Vollmer B: Effects of α-aminooxy-β-phenylpropionic acid on phenylalanine metabolism in p-fluorophenylalanine sensitive and resistant tobacco cells. Z Naturforsch 34C:770, 1979.
118. Gathercole RWE, Street HE: Isolation, stability and biochemistry of a p-fluoro-phenylalanine-resistant cell line of *Acer pseudoplatanus* L. New Phytol 77:29, 1979.
119. Bressan RA, Hasegawa PM, Handa AK: Resistance of cultured higher plant cells to polyethylene glycol-induced water stress. Plant Sci Lett 21:23, 1981.
120. Schiel O, Martin B, Piehl G-W, Nowak J, Hammer J, Sasse F, Schaer W, Lehmann J, Berlin J: Some technological aspects on the production of cinnamoyl putrescines by cell suspension cultures of *Nicotiana tabacum*. In "Third European Congress on Biotechnology." Weinheim: Verlag Chemie, 1984, Volume 1, p 167.
121. Bevan M: Binary *Agrobacterium* vectors for plant transformation. Nucleic Acids Res 12:8711, 1984.
122. Lörz H, Baker B, Schell J: Gene transfer to cereal cells mediated by protoplast transformation. Mol Gen Genet 199:178, 1985.
123. Wink M, Witte L: Turnover and transport of quinolizidine alkaloids. Diurnal fluctuations of lupanine in the phloem sap, leaves and fruits of *Lupinus albus* L. Planta 161:519, 1984.
124. Noé W, Berlin J: Induction of de-novo synthesis of tryptophan decarboxylase in cell suspensions of *Catharanthus roseus*. Planta 166:500, 1985.
125. Martin SM: Mass culture systems for plant cell suspensions. In Staba EJ (ed): "Plant Tissue Culture as a Source of Biochemicals." Boca Raton, FL: CRC Press, 1980, p 149.
126. Noguchi M, Matsumoto T, Hirata Y, Yamamoto K, Katsuyama A, Kato A, Azechi S, Kato K: Improvement of growth rates of plant cell cultures. In Barz W, Reinhard E, Zenk MH (eds): "Plant Tissue Culture and Its Biotechnological Application." Berlin: Springer-Verlag, 1977, p 85.
127. Tanaka H: Technological problems in cultivation of plant cells at high density. Biotechnol Bioeng 23:1203, 1981.
128. Wagner F, Vogelmann H: Cultivation of plant tissue cultures in bioreactors and formation

of secondary products. In Barz W, Reinhard E, Zenk MH (eds): "Plant Tissue Culture and Its Biotechnological Application." Berlin: Springer-Verlag, 1977, p 245.

129. Kiese S, Ebner HG, Onken U: A simple laboratory airlift fermentor. Biotechnol Lett 2:345, 1980.

130. Dainty AL, Goulding KH, Robinson PK, Simpkins I, Trevan MD: Effect of immobilization on plant cell physiology—real or imaginary? Trends Biotechnol 3:59, 1985.

131. Loh VY, Richards SR, Richmond P: Particle suspension in a circulating bed fermenter. Chem Eng J 32:B39, 1986.

132. Rosevear A: Biologically-active composites. UK Patent Application GB 2096169 A; 1982.

133. Nakagawa K, Konagai A, Fukui H, Tabata M: Release and crystallization of berberine in the liquid medium of *Thalictrum minus* cell suspension cultures. Plant Cell Rep 3:254, 1984.

134. Sato F, Yamada Y: High berberine-producing cultures of *Coptis japonica* cells. Phytochemistry 23:281, 1984.

135. Renaudin J-P: Uptake and accumulation of an indole alkaloid, [^{14}C] tabernanthine, by cell suspension cultures of *Catharanthus roseus* (L) G. Don and *Acer pseudoplatanus* L. Plant Sci Lett 22:59, 1981.

136. Renaudin J-P, Guern J: Compartmentation mechanisms of indole alkaloids in cell suspension cultures of *Catharanthus roseus*. Physiol Vég 20:533, 1982.

137. Deus-Neumann B, Zenk MH: A highly selective alkaloid uptake system in vacuoles of higher plants. Planta 162:250, 1984.

138. Deus-Neumann B, Zenk MH: Accumulation of alkaloids in plant vacuoles does not involve an ion-trap mechanism. Planta 167:44, 1986.

139. Parr AJ, Robins RJ, Rhodes MJC: Release of secondary products by plant cell cultures. In Webb C, Mavituna F (eds): "Process Possibilities for Plant and Animal Cell Cultures," I Chem E Symposium, UMIST 1986. Chichester: Ellis Horwood Ltd., 1987, p 229.

140. Ojima K, Abe H, Ohira K: Release of citric acid into the medium by aluminium-tolerant carrot cells. Plant Cell Physiol 25:855, 1984.

141. Messens E, Lenaerts A, Van Montagu M, Hedges RW: Genetic basis for opine secretion from crown gall tumour cells. Mol Gen Genet 199:344, 1985.

142. Ojima K, Ohira K: Nutritional requirements of callus and cell suspension cultures. In Thorpe TA (ed): "Frontiers of Plant Tissue Culture 1978." University of Calgary: International Association for Plant Tissue Culture, 1978, p 265.

143. Berlin J: The use of immobilized plant cells—an evaluation. Int Assoc Plant Tissue Culture Newsletter 46:8, 1985.

144. Robins RJ, Rhodes MJC: The stimulation of anthraquinone production by *Cinchona ledgeriana* cultures with polymeric adsorbents. Appl Microbiol Biotechnol 24:35, 1986.

145. Wiermann R: Secondary plant products and cell and tissue differentiation. In Conn EE (ed): "The Biochemistry of Plants—A Comprehensive Treatise. Volume 7-Secondary Plant Products." New York: Academic Press, 1981, p 85.

146. Boudet AM, Alibert G, Marigo G: Vacuoles and tonoplast in the regulation of cellular metabolism. In Boudet AM, Alibert G, Marigo G, Lea PJ (eds): "Membranes and Compartmentation in the Regulation of Plant Functions," Ann Proc Phytochem Soc Eur, Vol 24. Oxford: Oxford University Press, 1984, p 29.

147. Felix H: Permeabilized cells. Anal Biochem 120:211, 1982.

148. Brodelius P, Nilsson K: Permeabilization of immobilized plant cells, resulting in release

of intracellularly stored products with preserved cell viability. Eur J Appl Microbiol Biotechnol 17:275, 1983.

149. Tanaka H, Hirao C, Semba H, Tozawa Y, Ohmomo S: Release of intracellularly stored 5′−phosphodiesterase with preserved plant cell viability. Biotechnol Bioeng 27:890, 1985.

150. Parr AJ, Robins RJ, Rhodes MJC: Permeabilization of *Cinchona ledgeriana* cells by dimethylsulphoxide. Effects on alkaloid release and long-term membrane integrity. Plant Cell Rep 3:262, 1984.

151. Rueffer M: The production of isoquinoline alkaloids by plant cell cultures. In Phillipson JD, Roberts MF, Zenk MH (eds): "The Chemistry and Biology of Isoquinoline Alkaloids." Berlin: Springer-Verlag, 1985, p 265.

152. Kurkdjian A, Barbier-Brygoo H, Manigault J, Manigault P: Distribution of vacuolar pH values within populations of cells, protoplasts and vacuoles isolated from suspension cultures and plant tissues. Physiol Vég 22:193, 1984.

153. Roberts JKM, Ray PM, Wade-Jardetzky N, Jardetzky O: Estimation of cytoplasmic and vacuolar pH in higher plant cells by ^{31}P NMR. Nature 283:870, 1980.

154. Matile Ph: Biochemistry and function of vacuoles. Annu Rev Plant Physiol 29:193, 1978.

155. Rataboul P, Alibert G, Boller T, Boudet AM: Intracellular transport and vacuolar accumulation of o-coumaric acid glucoside in *Melilotus alba* mesophyll cell protoplasts. Biochim Biophys Acta 816:25, 1985.

156. Marty F, Branton D, Leigh RA: Plant vacuoles. In Tolbert NE (ed): "The Biochemistry of Plants—A Comprehensive Treatise. Volume 1—The Plant Cell." New York: Academic Press, 1980, p 625.

157. Wittenbach VA, Lin W, Hebert RR: Vacuolar localization of proteases and degradation of chloroplasts in mesophyll protoplasts from senescing primary wheat leaves. Plant Physiol 69:98, 1982.

Biotechnology in Agriculture, pages 35–52
© 1988 Alan R. Liss, Inc.

Growth Control in Plant Tissue Cultures

Andrew Goldsworthy

Department of Pure and Applied Biology, Imperial College,
London SW7 2BB, England

————◆◆————

————◆◆————

Plant tissue cultures can do some very useful things. Isolated shoots are used for the micropropagation of plant clones. Some tissue and cell cultures can produce valuable secondary products, and protoplast cultures can be used for genetic engineering. But to do these things effectively, we must control the growth of tissue cultures and direct it along channels of our own choosing, either towards secondary product synthesis or towards plantlet regeneration.

So far, our success has been limited. Few cultures produce secondary products in commercially significant amounts [1]; not all regenerate plantlets; many do so only with difficulty [2]; and the resulting plants often show genetic aberrations [3]. Perhaps our greatest failing is that most of our knowledge is based on empirical studies with as yet little understanding of the mechanisms controlling growth and differentiation.

In this chapter, I will not attempt to catalog the culture methods for growth and regeneration in specific tissues, since this could occupy many volumes, and they have been well reviewed elsewhere [4–6]. Instead, I shall try to draw together what we know of some of the underlying mechanisms and use it to create an overview which might form a basis for further research. This will include a brief outline of the way in which chemical signals may control growth and differentiation, and a more detailed account of the recently discovered and remarkable effects of weak electric currents. Where possible, I will adopt an evolutionary approach, since an attempt to trace evolutionary pathways frequently makes sense of seemingly unrelated and inexplicable empirical data.

I. DIFFERENTIATION

Perhaps the most important aspect of the behavior of tissue cultures is the phenomenon of differentiation, since this is vital for the regeneration of plantlets from genetically engineered cells and may also enhance the formation of secondary products in cultures where these normally accumulate in differentiated structures.

Differentiation may be defined as the specialization of different regions of the plant to perform specific tasks, allowing each to be carried out more efficiently. To understand it, we need to know the mechanisms by which the different regions become specialized and also how they regulate one another's activities for the plant to function as an integrated whole. We are only just beginning to learn how they do these things, but from what we have learned so far, they appear to use a mixture of chemical and electrical control signals.

II. CHEMICAL CONTROL SIGNALS

A. The Evolution of Chemical Signals

Even primitive and seemingly undifferentiated unicells have the seeds of differentiation. For example, they possess specialized organelles; and the simplest eukaryotic cell will have a nucleus, mitochondria and perhaps chloroplasts, all performing different functions but regulating one another's activity to give a viable organism. Each system interacts with the others, largely by chemical means, such as by providing essential substrates for one another or by the allosteric regulation of each other's enzymes.

From these simple beginnings, evolution led to unicells showing true differentiation where whole regions of the organism became specialized for different functions. An example of this is *Acetabularia,* which is an umbrella-shaped alga consisting of a large and complex photosynthetic cap on a slender stalk anchored to the substrate by a rhizoid. These regions are interdependent—the cap provides most of the food; the stalk must be long enough and strong enough to hold it above competing organisms; and the rhizoid, as well as anchoring it to the substrate, provides a protected environment for the nucleus. The molecular mechanisms by which *Acetabularia* becomes differentiated lie in the realms of electrophysiology, and I will return to this subject later. Let us first look briefly at its chemical control systems. A surprising discovery from work with this organism was that a small piece of stalk from which the nucleus had been removed could regenerate into a fully differentiated alga [7]. In *Acetabularia* at least, changes in the pattern of gene transcription do not play an important part in either differentiation or the coordination of metabolism in the different regions. Presumably, many of the chemical control systems which evolved to regulate metabolism in undifferentiated organisms are still functional, albeit with longer diffusion pathways as the differentiated regions are now spacially separated. In particular, we would expect the substances coordinating metabolism still to be normal metabolites or their derivatives, and like most intermediary metabolites, they will not easily leak out through the cell membrane. This is an important point with respect to the further evolution of chemical control signals and to their biotechnological applications, and I will return to it later.

Evolution from the unicell led to larger multinucleate structures like the coenocytes of many filamentous algae. Differentiation, e.g., into photosynthetic and nonphotosynthetic regions, could still occur, and the lack of cross-walls still allowed food and control signals to pass from one part to another. Further evolution led to the subdivision of the plant body into "cells," each with a single nucleus. But total compartmentalization into units *completely* separated by membranes would have been a major impediment to the transport of both food and control signals. Nature avoided this by making compartmentalization incomplete. The "cells" remained connected either by relatively large cytoplasmic "pits" in their cell walls as in some present-day algae, or by tens of thousands of extremely thin cytoplasmic "plasmodesmata" as in other algae and the higher plants. The interior of groups of cells placed in cytoplasmic continuity in this way forms what we now term the *symplast,* and the region outside the symplast (mainly the cell walls) forms the *apoplast.*

The cytoplasmic connections between cells allowed the existing substrates and control signals to be passed from one cell to the next, and even newly evolved signals could be transported in the same simple way. Signals transmitted by secretion into the apoplast followed by reabsorption by neighboring cells would be less likely to evolve at this stage since evolution to the cellular condition occurred in an aquatic environment where such substances would be lost to the surrounding medium. It is therefore likely that many of the substances coordinating the function of neighboring cells even in higher plants are still normal metabolites traveling mainly in the symplast. For this reason their identification may be very difficult, and this may explain why we know so little about them.

Compartmentalization into cells added another dimension to differentiation. The location of each nucleus in its own "private" cytoplasm allowed different cells to be programmed to express only those parts of their genome necessary for their specialist functions. But to do this each cell must have information about its position in the plant in the form of chemicals capable of inducing or repressing the synthesis of specific enzymes. This information has to come from other parts of the plant. We only know the identity of a few of these substances, but many of them may also be normal metabolites whose concentration just happens to give the cell a clue to its neighbors. For example, developing vascular bundles contain high concentrations of sucrose in the phloem and indole-3-acetic acid (IAA) in the xylem; and there is evidence, both from whole plants and tissue cultures [8], that a mixture of these substances stimulates and controls the differentiation of new vascular tissue as the bundles extend. In particular, the sucrose concentration appears to increase the proportion of phloem formed [9], perhaps as a means of matching the amount of phloem to the amount of sugar needing translocation. The use of normal metabolites in this way to control differentiation is not surprising because natural selection will favor organisms whose cells respond to chemical clues *which already exist.* "Specially invented" signals are less likely because selection for signal production cannot occur in the absence of an advantageous response and selection for the response cannot occur in the absence of the signal.

The chemical signals controlling differentiation are divided, arbitrarily, into *morphogens,* which carry information over short distances and are responsible for the precise positioning of individual differentiated cells, and *growth hormones,* which carry information over longer distances and are responsible for the grand strategy of growth. There is, however, no clear distinction between the two, and some substances such as IAA may function as both [8].

B. Morphogens and Cell Differentiation

Morphogens are substances whose existence remains largely hypothetical but which are believed to control the differentiation of individual cells and their ordered arrangement into tissues and organs [8]. It would be valuable to know more about them for, given the right mix of morphogens, it might be possible to produce a tissue culture containing only one kind of cell (say gland cells) and so increase the in vitro yield of secondary products.

Unfortunately, little is known about the morphogens; there could be hundreds of them, and they could be anything from complex organic compounds to simple metal ions. Because each may only be formed in a few cells and affect just their immediate neighbors, it is very difficult to extract them in sufficient quantities for identification. Even then, identification might still be difficult because, if (as seems likely) they are transmitted mainly in the symplast and do not penetrate cell membranes, their biological effects will not be detected by supplying them externally. Much of the evidence for their existence comes instead from surgical and grafting experiments [8].

Many different morphogens may be needed to produce the extremely complicated tissue patterns found in the intact plant. Even individual cells may need information from more than one source to get a precise "fix" on their position in three dimensions before they will differentiate. One way they appear to obtain this information is from the *balance* between two or more morphogens coming from different directions. A possible example is the differentiation of higher plant cambium. This is normally laid down between the xylem and phloem, and there is evidence that it may be formed at a point on two opposing concentration gradients (one for IAA leaving the xylem and the other for sucrose leaving the phloem) where the two substances are in a predetermined ratio [8].

A practical conclusion which might be drawn from these observations is that although it may be theoretically possible to produce a tissue culture consisting only of useful cells such as gland cells by the application of morphogens, in practice it may be extremely difficult. The necessary morphogens may be hard to identify; a precise balance between several may be necessary; and if even only a proportion of them do not penetrate cell membranes, it may be impossible to obtain the desired effect by simply adding them to the culture medium. It is perhaps for this reason that most of our success in controlling differentiation in plant tissue cultures has come from the use of hormones rather than morphogens.

C. Growth Hormones and Strategic Control

Chemical signals operating over longer distances became necessary as plants became larger, and this resulted in the evolution of *hormones*. Unlike the putative morphogens, the known plant hormones can easily cross membranes, which they must do to enter translocatory tissue such as the xylem and phloem (even the polar transport of IAA which occurs in apparently ordinary cells involves its crossing the membrane of each cell along the way [10]). They achieve this either by having a high lipid solubility enabling them to diffuse through the lipid phase of the membrane or (as with IAA) by having a specific transmembrane transport systems. Either way, this makes it relatively easy to demonstrate their biological effects simply by adding them externally to cells and tissues, and this may explain why we know much more about hormones than we do about morphogens.

As with the morphogens, it is often the balance between hormones rather than their absolute concentrations which produces the biological effect, but unlike the morphogens, which give precise positional information for individual cells, the main function of growth hormones seems to be in controlling the grand strategy of growth for large areas of the plant. An important example of this is the auxin/cytokinin ratio which determines whether a damaged plant or a tissue culture regenerates either roots or buds.

D. Mechanisms of Action

Very little is known of how most morphogens and hormones work, but the fact that in either case the balance between them is often more important than their absolute concentrations suggests that there may be similarities. Perhaps the growth hormones are simply long-range morphogens and function and interact by similar mechanisms. This is supported by the observation that IAA can behave as both a morphogen and a hormone [8]. The effects of IAA have, of course, been well studied and include changes in the pattern of gene transcription and also stimulations of hydrogen ion excretion promoting cell-wall elongation [11] and perhaps also promoting the uptake of nutrients by ion cotransport. If the action of other morphogens and hormones share this degree of complexity, the chemical control of differentiation could be a very complex process indeed. Nevertheless, despite our lack of detailed knowledge, we have been able to use plant hormones in a variety of applications, including the regeneration of plants from tissue cultures.

III. THE REGENERATION OF PLANTLETS

The regeneration of organs or plantlets from tissue cultures is an essential step in the recovery of whole plants from genetically manipulated cells, and it is usually achieved by making the culture conditions conducive to regeneration by quasinatural mechanisms. In nature, whole new plants are regenerated from the zygotic embryo in the seed, or specific organs may be regenerated from vegetative parts to replace those lost by accidental damage. These phenomena have their in vitro equivalents, namely, somatic embryogenesis and organogenesis.

A. Somatic Embryogenesis

Somatic embryogenesis is the formation from somatic cells of embryolike structures capable of growing into new plants. It is relatively rare in intact plants, but it can often be induced artificially in tissue cultures [12]. Embryogenesis can occur in either callus or suspension cultures, but the cells must be as close as possible to the zygotic condition and show little or no differentiation. The most suitable explants come from undifferentiated embryonic or meristematic tissues, but even then, hormone treatment is usually needed to help them dedifferentiate further.

Embryogenic cultures are usually produced by an initial treatment with high auxin concentrations which presumably swamp the natural control mechanisms which would otherwise lead to differentiation. The best effects are usually obtained with synthetic auxins such as 2,4-dichlorophenoxyacetic acid (2,4-D), which are more potent and not biodegradable. However, the high auxin concentrations needed to induce the dedifferentiation of cells to the zygotelike state also prevent their redifferentiation to form embryos. To allow this to occur, the cultures must be transferred to a medium containing little or no auxin. Recognizable embryos may then be formed, eventually giving rise to plantlets which can be transferred to soil.

B. Organogenesis

In nature, this is how damaged plants regenerate lost parts; for example, a shoot cutting taken from a plant can regenerate new roots, and adventitious buds may grow either directly on the remaining stump or on callus formed over its cut surface. Natural organogenesis is under hormonal control, and the main signal controlling whether buds or roots develop is the auxin/cyto-kinin ratio [13]. Auxin is produced near the shoot apex and cytokinins in the roots. The loss of the root system increases the auxin/cytokinin ratio and stimulates the formation of new roots, whereas the loss of the shoot apex reduces the auxin/cytokinin ratio and stimulates the growth of buds.

The same mechanism controls organogenesis in tissue cultures—for example, either roots or shoots can be induced directly on differentiated explants by culturing them on a medium containing *low* concentrations of auxin and cytokinin in the appropriate ratio. Higher concentrations suppress differentiation and usually cause the formation of a callus instead, but if the concentrations are subsequently lowered, organogenesis may then occur on the callus itself. This phenomenon was discovered by Skoog and Miller [14] and applies to most species which have been examined. The exact concentrations of auxin and cytokinin required vary with species, and sometimes additional growth promoters may be helpful [2]. This technique has been used extensively for the regeneration of plantlets from callus by first inducing shoots and then changing the auxin/cytokinin ratio to induce rooting. However, even when the best rates of organogenesis have been achieved by adjusting the medium, the results often fall short of what we would like. Could there be other factors still missing? One such factor appears to be a sense of polarity.

IV. POLARITY

Nearly all plants show polarity; i.e., one end differs from the other. For example, in higher plants there is a root at one end and a shoot at the other. Polarity is a property of the individual cells from which the plant is made, and each cell ''knows'' which is its top and which is its bottom. This can be seen in the polar transport of IAA, which is normally translocated from the shoot apex towards the roots, but small sections of shoot will still drive it in this direction even when they are removed from the rest of the plant and turned upside down [10].

A. The Electrical Control of Polarity

How does a plant cell ''know'' one end of itself from the other, and how is this information translated into the direction of growth? The first clue to this came when Elmer Lund [15] passed an electric current through seawater-containing *Fucus* eggs and found that they germinated with their rhizoids pointing towards the positive electrode, i.e., the current was controlling their polarity. This led to the speculation that natural electric currents generated by the cells themselves may normally control polarity, and just over 40 years later, Lionel Jaffe discovered them them, also in *Fucus* eggs [16]. *Fucus* eggs are symmetrical spheres, and the direction in which they germinate is normally determined by the direction of light with the rhizoid, which normally anchors it to the substrate, growing from the shaded side. Jaffe was able to show that unidirectional illumination caused an electric current to flow through the eggs from the shaded to the illuminated side several hours before any growth occurred, suggesting that this current might be responsible for the orientation of the polar growth which followed.

He and Richard Nuccitelli then went on to develop the *vibrating probe* [17], which is an instrument for mapping the tiny electric currents flowing through individual cells by measuring the voltage gradients in the surrounding medium where they enter and leave. Using this and other methods, natural currents flowing with densities on the order of microamps per square centimeter have been measured flowing through the cells of both animals and plants at all levels of evolution in a direction corresponding to their morphological polarity [18–20]. Also, there are several reports of artificially applied currents controlling the direction of polar growth of other cells such as regenerating *Acetabularia* stalks [21] and the spores of *Funaria* [22] and *Neurospora* [23]. Taken together, this provides compelling evidence that natural electric currents control cellular polarity.

B. The Generation of Natural Currents

In single cells, the currents flow in one direction through the interior and complete the circuit in the external medium. They are carried by ions and are generated by metabolically driven ion pumps in the cell membrane similar to those used for the active uptake of mineral nutrients. Having been driven in one direction through the membrane by the pumps, the ions return by passive but specific ion channels. The pumps and/or channels are asymmetrically distributed over the cell surface so that the ion species in question tends to enter and leave at opposite ends, causing an electric current to flow through the cell [18,20].

The ions responsible have been determined, either by using radiotracers or by measuring the change in current flow when specific ions are omitted from the external medium. In general, the results indicate that many different ion species contribute to the current, with their relative contributions varying with the biological material and its physiological condition. In many cases, hydrogen ions carry a large proportion of the current, but there are also significant contributions by metal ions such as calcium, sodium, and potassium as well as anions such as chloride [20].

The mechanism by which the current is first initiated in hitherto nonpolar cells has been studied in the germinating fucoid egg [24]. Much of the current immediately following unilateral illumination is carried by calcium, but once it has begun to flow, the current grows steadily and the involvement of other ion species increases, even if the unidirectional light stimulus is removed. It is suggested that the primary effect of unidirectional illumination may be to open a small number of calcium channels on the shaded side, allowing a weak current carried by calcium ions to drawn into the negatively charged interior of the cell in this region, this internal charge being maintained by more uniformly distributed ion pumps. But after this, a positive feedback mechanism comes into operation which magnifies the current and stabilizes the cell's polarity by increasing both the number and variety of ion channels and pumps functioning in the regions where the current already enters and leaves.

C. The Evolution of Transcellular Currents

Although at first sight, the electrical control of polarity seems both bizarre and unlikely, it is not difficult to see how it may have evolved. Electrical control of cell polarity has so far only been demonstrated in eukaryotes [25], so let us consider a primitive, more-or-less symmetrical eukaryotic cell living heterotrophically on an organic substrate. The active uptake of nutrients from that substrate would presumably be by ion cotransport, probably with hydrogen ions which had been actively excreted by the cell. Since hydrogen ion excretion would be taking place all over the cell, but cotransport would be concentrated in the region next to the food source, there would be a net flow of current carried by hydrogen ions through the cell from its point of contact with the substrate. Even today, much of the transcellular current determining the polarity of fungal hyphae can enter by hydrogen ion cotransport with organic substrates at the hyphal tip [26]. Given the existence of such a current in our primitive eukaryote, natural selection would favor any mechanism which concentrated growth at its point of entry so that the cell could grow more deeply into the nutrient substrate. It would also favor any positive feedback mechanisms which increased the strength of the current and made it more effective. This then may be the way in which elongated, tip-growing polar structures such as fungal hyphae first evolved.

The evolution of photosynthetic eukaryotes probably came later after the acquisition of photosynthetic endosymbionts [27]. Growth towards the substrate was no longer important; growth now had to be towards the light. The switch-over may have been simple. All that was necessary was the evolution of a light-sensitive pigment in the membrane which could open or close ion channels according to the light intensity. This would result in an asymmetric distribution of open channels between the illuminated and shaded side and generate a current along the axis of illumination. The organism could then grow towards the light by the existing electrophysiological mechanisms.

D. The Mechanisms of Action

The mechanisms by which transcellular electric currents control polarity are not fully understood, but there are several possibilities. For example, calcium ions tend to enter at the growing tips of cells and may activate enzymes to cause growth in these regions [28], or the currents could bring about the electrophoretic transport of charged metabolites, or charged vesicles containing substrates, to the growing cell tips [24]. Another and perhaps more intriguing possibility is that transcellular currents distribute different proteins electrophoretically along the length of the cell according to their charge. Work on animal cells has shown that proteins can move laterally within the plane of the membrane in electrical fields of the same order as those occurring in nature [29]. The more highly charged ones would tend to accumulate at the electrical poles of the cell and displace those less highly charged to equatorial positions. Thus the cell's DNA would not only control a protein's enzyme activity by determining its amino acid sequence, it could also control its exact position along the cell's axis by determining its amino acid balance and hence its overall charge. Such an asymmetric distribution of proteins, either in the membrane or in those parts of the cytoplasm unaffected by streaming could cause metabolic asymmetries leading to changes in the cytoskeleton and precise patterns of differentiation within the cell including physiological and morphological polarity.

Such a mechanism could also control the formation of differentiated multicellular structures; for example, proteins determining the plane of cell division might be positioned at the correct location along the cell's axis in this way. It also provides a means by which cell division can result in nonidentical daughter cells. For example, when a meristematic cell divides, it normally gives one cell which remains meristematic and another which matures and differentiates. This could be because the cytoplasms of each daughter are electrophoretically different and behave differently, perhaps activating different sets of genes.

E. Natural Currents in Multicellular Structures

Weak electric currents corresponding to the direction of growth have been detected in many multicellular structures such as algal filaments, cereal coleoptiles, and isolated embryos growing in culture. The filamentous alga *Pithophora* grows by the repeated division of an apical cell, but all the cells have the same electrical polarity, and changes in the pattern of current flow predict the location of the apex of each new daughter cell before it is cut off

[30]. In the cereal coleoptile, the pattern of current flow is a little more complex but is predominantly longitudinal with the apex negative to the base as measured by externally placed electrodes [31]. Carrot [32] and tobacco [33] embryos also have their apical ends negative to the base, and the density of the current flowing through them increases severalfold as they leave the globular stage and become visibly polar. Perhaps these currents, which now follow defined multicellular pathways, coordinate the polarities of the cells so that they can grow into organized polar structures.

F. Polarity in Callus Cultures

When we look at an undifferentiated callus culture, we see little evidence for a coordination of cellular polarities. There are no apical meristems; the overall shape is amorphous; and growth is usually by small and disorganized regions of dividing cells scattered throughout the tissue. Any vascular structures present tend also to be disorganized, often forming spherical nodules of xylem surrounded by phloem [8], as if these cells had the information to differentiate in the right relative positions (probably provided by locally generated morphogens) but none on the direction for elongation. Even when true vascular strands are produced, they have no consistent orientation. It seems that the cells of undifferentiated tissue cultures are deficient in the signals needed to coordinate their polarities, perhaps because they lack an organized flow of electric current. The question we sought to answer was, What effect would it have on callus growth and differentiation if an electric current were to be applied artificially?

V. EXPERIMENTS WITH ARTIFICIAL CURRENTS IN TISSUE CULTURES

A. The Biological Material

Callus cultures were initiated from hypocotyl sections of aseptically grown seedlings of *Nicotiana tabacum* var. *virginica*. Each experiment was performed on a single clone after three passages of 4–5 wk under weak illumination in Gamborg's B5 medium [34], pH 5.5, containing 0.1 ppm kinetin and 1 ppm 2,4-D to suppress differentiation.

B. Electrical Treatment

The first problem to overcome was the polarization of the electrodes. Metal electrodes usually polarize owing to the formation of electrolytic products at their surface which can either generate a reverse voltage or form an insulating coating over them. In either case, the current flow through the tissue becomes reduced. This effect was minimized by supplying the current

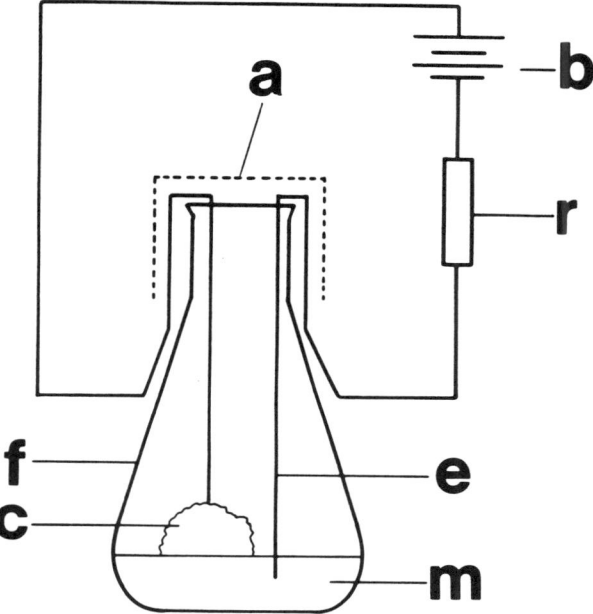

Fig. 1. An inexpensive apparatus for the electrical treatment of callus cultures. The current is derived from a moderately high voltage source via a high-value resistor. Several hundred cultures can be connected to the same battery, but each must have its own resistor to limit the current. The arrangement shown will provide approximately 1.8 μA for each culture: a, aluminum foil cap; b, 18-V battery; c, callus; e, stainless steel electrode; f, flask; m, agar medium; r, 10-MΩ resistor.

from a moderately high-voltage source via a high-value resistor. A diagram of the apparatus is shown in Figure 1. Each culture flask has its own resistor, but they can all share a common power supply. Provided the supply voltage is much greater than the reverse voltages owing to polarization and that the value of the resistor is much higher than the resistance of the rest of the circuit, the supply voltage and the resistor will be the main factors determining the current flow. The approximate current flow through the tissue can then be calculated from these values by Ohm's law—for example, a 10-MΩ resistor with a 20-V power supply gives approximately 2 μA. The power supply can, if necessary, simply be a pair of 9-V transistor radio batteries connected in series to give 18 V and therefore 1.8 μA through a 10-MΩ resistor. Because the current drain is so low, one set of batteries should supply hundreds of cultures for the likely duration of any normal experiment.

The electrodes were made from thin stainless steel wire (uninsulated diameter 0.25 mm) insulated with Teflon to prevent short circuits where they passed under the aluminum, foil cover of the culture flasks. Suitable wire can be obtained from from Clark Electromedical Ltd. (Pangbourne, England). Each electrode consisted of a 15-cm length of this wire bared for a few millimeters at either end and autoclaved before use.

The experimental cultures were electrically treated in 100-ml Erlenmyer flasks containing 25 ml of Gamborg's B5 agar medium with various supplements. They were inoculated with callus weighing about 160 mg. One electrode was inserted aseptically for about 2 mm into the top of the callus, and the other inserted about 20 mm away in the culture medium. Currents of either 1 or 2 μA were driven continuously between them for the duration of the experiment. The objective was to make the current "fan out" to give a range of densities as it passed from the electrode in the callus to the oppositely charged culture medium, for example, 1 μA would give current densities from about 60 $\mu A \cdot cm^{-2}$ immediately around the electrode to about 0.2 $\mu A \cdot cm^{-2}$ at the base when the callus was fully grown. Assuming a tissue resistivity of 5,000Ω · cm, this represents voltage gradients between 300 and 1 $mV \cdot cm^{-1}$. Thus, if any physiological effect was critically dependent on either current density or voltage gradient within these ranges, suitable conditions should occur somewhere within each callus.

C. The Stimulation of Shoot Formation

When the cultures were put onto media conducive to shoot formation, these organs began forming after about 3 wk (which is normal for tobacco), but about five times as many were produced in the electrically treated cultures as in the controls where the electrodes were not connected to the power supply [35]. The first shoots tended to be formed in the more negative regions of the callus, implying some polarization of the whole tissue; but eventually they formed all over the cultures, suggesting that the exact current density needed was not very critical within the range tested.

The stimulation of organogenesis occurred with currents of either polarity and so could not be attributed to the electrophoretic movement of growth-promoting substances from the culture medium or their electrolytic generation at the electrode in the callus. It could, however, be explained if we postulate that the electric current aligned the polarities of the callus cells predominantly in the same direction, allowing more of the dividing cells to become organized into shoot-forming meristems.

D. The Stimulation of Callus Growth

Further evidence for the electrical control of tissue culture polarity came from similar experiments using culture media not conducive to differentiation [36,37]. Here, the current caused a 70% stimulation of growth instead, but unlike the effect on organogenesis, it only occurred with current in one direction, with perhaps a slight inhibition when it was passed the other way. The electrical effect was also dependent on there being IAA in the medium and was inhibited by 2,3,5-tri-iodobenzoic acid (TIBA), which is an inhibitor of the polar transport of IAA [38]. The effect also disappeared when the IAA was replaced by 2,4-D, which is an auxin showing little or no polar transport [10]. These findings are also consistent with the hypothesis that the electric current had coordinated the polarities of the cells. When polarized in one direction, their polar transport mechanisms would draw in auxin from the medium to stimulate growth, but when polarized in the reverse direction, auxin would tend to be expelled, and growth would be inhibited. The electrical effect would not be expected when the polar transport of IAA was inhibited or with auxins not showing polar transport.

E. The Mechanism of Cell Alignment

It is unlikely that the externally applied current mimics the natural currents by actually flowing through the cells because the electrical resistance of cell membranes is extremely high. Instead it must flow mostly in the cell walls where it cannot affect internal metabolism directly. Nevertheless, it could still influence cellular polarity if it brings about the electrophoretic redistribution of the ion channels and pumps in the cell membrane to align the cell's own current with the one artificially applied.

Since the currents we applied were of the same order as those reported to flow naturally, it seems likely that nature may also use this mechanism to coordinate cellular polarities. For example, if the ion channels and pumps making one end of a cell negative were themselves negatively charged, they would be attracted electrophoretically to the positive end of the neighboring cell. In this way, the electrical axes of the cells making up a tissue would tend to align themselves nose to tail in parallel rows like iron filings in a magnetic field. The initial polarity would be established when each cell was cut off from a polar meristematic cell; but thereafter they would tend to hold eachother in stable rows.

This interpretation is also consistent with observations made by Webster and Schrank [39], who applied transverse currents to cereal coleoptiles. They too found that auxin was apparently drawn towards the negative electrode. Perhaps this was due to the applied current deflecting the normal vertical polarity of the cells. In nature, the base of the coleoptile is normally electropositive to its apex, implying that the basal ends of its component cells are also electropositive to their apices. On our hypothesis, the positive basal end of each cell would align itself next to the negative apical end of the cell beneath *or* towards the negative electrode in an artificially applied field. The normal basipetal transport of auxin would therefore be deflected towards the negative electrode when an external electric current is applied.

We can therefore explain both the stimulation of growth and the stimulation of organogenesis in tissue cultures in terms of a realignment of their cellular polarities with their "basal" ends apparently directed towards the negative electrode. Presumably, the initial alignment was lost when the callus was induced by rapid cell division in the explant away from the stabilizing electrical influence of the rest of the plant. The artificially applied current may simply restore this alignment to make the natural processes of organ regeneration more likely.

VI. POSSIBLE APPLICATIONS

A. Enhanced Plantlet Regeneration

The possible use of the technique in assisting the regeneration of plantlets by organogenesis from genetically engineered cell colonies is perhaps self-evident. Not all tissue cultures regenerate easily [2], and cells which have been genetically manipulated in vitro may be particularly recalcitrant because of possible genetic imbalance and the physiological battering they may have received in the process. If the production of plantlets is limited by the ability of the manipulated cultures to regenerate, a severalfold increase by the application of electric currents would be very welcome. If, as we have found, it results in an approximately fivefold stimulation of organogenesis, this may mean a fivefold reduction in the number of experiments in genetic manipulation needed and a corresponding reduction in costs. Since such experiments tend to be rather expensive, and electrical treatment is cheap (the additional cost in materials is pence per culture flask), there could be a considerable financial saving.

B. Increased Genetic Stability

The genetic instability of undifferentiated tissue cultures has been attributed to aberrations during the division of cells not forming part of organized meristems [40]. If electrical treatment increases the degree of organization within the cultures, it may also enhance genetic stability. This could make callus cultures more suitable for routine commercial plant propagation and might even provide a means for the long-term maintenance of germplasm.

C. Enhanced Secondary Product Synthesis

Similar electrical treatment of immobilized cultures could enhance the synthesis of secondary products, both by increasing growth or and by stimulating the formation of differentiated structures in which such products might accumulate. No major experimental work has yet been done on this topic, but it seems a worthwhile area of investigation. If useful effects are discovered, the problem of scaling the process up for industrial application could be a challenging and fascinating test of ingenuity.

REFERENCES

1. Fowler MW: Industrial applications of plant cell culture. In Yeoman MM (ed): "Plant Cell Culture Technology." Oxford: Blackwell, 1986, p 202.
2. Flick CE, Evans DA, Sharp WR: Organogenesis. In Evans DA, Sharp WR, Ammirato PV, Yamada Y (eds): "Handbook of Plant Cell Culture." Volume 1. New York: Macmillan, 1983, p 13.
3. Chaleff RF: "Genetics of Higher Plants: Applications of Cell Culture." Cambridge: Cambridge University, 1981.
4. Withers LA, Alderson PG (eds): "Plant Tissue Culture and its Agricultural Applications." London: Butterworths, 1986.
5. Dixon RA (eds): "Plant Cell Culture: A Practical Approach." Oxford: IRL, 1985.
6. Evans DA, Sharp WR, Ammirato PV, Yamada Y (eds): "Handbook of Plant Cell Culture." Volume 1. New York: Macmillan, 1983.
7. Kloppstech K: Acetabularia. In Grierson D, Smith H (eds): "The Molecular Biology of Plant Development." Oxford: Blackwell, 1982.
8. Warren Wilson J, Warren Wilson PM: Control of tissue patterns in normal development and in regeneration. In Barlow PW, Carr DJ (eds): "Positional Controls in Plant Development." Cambridge: Cambridge University Press, 1984, p 225.
9. Jeffs RA, Northcote DH: The influence of indol-3yl acetic acid and sugar on the pattern of induced differentiation in plant tissue cultures. J Cell Sci 2:77, 1967.
10. Goldsmith MHM: The polar transport of auxin. Annu Rev Plant Physiol 28:439, 1977.
11. Bandurski RS, Nonhebel HM: Auxins. In Wilkins MB (ed): "Advanced Plant Physiology." London: Pitman, 1984, p 1.
12. Tisserat B: Embryogenesis, organogenesis and plant regeneration. In Dixon RA (ed): "Plant Cell Culture: A Practical Approach." Oxford: IRL, 1985, p 79.
13. Wareing PF, Phillips IDJ: "Growth and Differentiation in Plants." Oxford: Pergamon, 1981.
14. Skoog F, Miller CO: Chemical regulation of growth and organ formation in plant tissue cultured in vitro. Symp Soc Exp Biol 11:118, 1957.
15. Lund EJ: Electrical control of organic polarity in the egg of *Fucus*. Bot Gaz 76:288, 1923.
16. Jaffe LF: Electrical currents through the developing *Fucus* egg. Proc Natl Acad Sci USA 56:1102, 1966.
17. Jaffe LF, Nuccitelli R: An ultrasensitive vibrating probe for measuring extracellular currents. J Cell Biol 63:614, 1974.
18. Jaffe LF, Nuccitelli R: Electrical controls of development. Annu Rev Biophys Bioeng 6:445, 1977.

52 ◇ Goldsworthy

19. Nuccitelli R: The involvement of transcellular ion currents and electric fields in pattern formation. In Malacinski GM, Bryant SV (eds): "Pattern Formation: a Primer in Developmental Biology." New York: Macmillan, 1984, p 23.
20. Nuccitelli R (ed): "Ionic Currents in Development." New York: Alan R. Liss, Inc., 1986.
21. Novak R, Bentrup FW: An electrophysiological study of regeneration in *Acetabularia mediterranea*. Planta Med 108:227, 1972.
22. Chen TH, Jaffe LF: Forced calcium entry and polarized growth in *Funaria* spores. Planta Med 144:401, 1979.
23. Gow NAR, McGillivray AN: Ion currents, electrical fields and the polarized growth of fungal hyphae. In Nuccitelli R (ed): "Ionic Currents in Development." New York: Alan R. Liss, Inc., 1986, p 81.
24. Jaffe LF, Robinson KR, Nuccitelli R: Local cation entry and self-electrophoresis as an intracellular localization mechanism. Ann NY Acad Sci 238:372, 1974.
25. Jaffe LF: Ion currents in development: an overview. In Nuccitelli R (ed): "Ionic Currents in Development." New York: Alan R. Liss, Inc., 1986, p 351.
26. Harold FM, Schreurs WJA, Caldwell JH: Transcellular currents in the water mold *Achlya*. In Nuccitelli R (ed): "Ionic Currents in Development." New York: Alan R. Liss, Inc., 1986, p 89.
27. Alberts B, Bray D, Lewis J, Raff M, Roberts K, Watson JD: "The Molecular Biology of the Cell." New York: Garland, 1983, p 541.
28. Brownlee C: Cytoplasmic free calcium in growing rhizoids of *Fucus serratus*. In Nuccitelli R (ed): "Ionic Currents in Development." New York: Alan R. Liss, Inc., 1986, p 71.
29. Poo M, Robinson KR: Electrophoresis of concanavalin A receptors along muscle cell membranes. Nature 265:602, 1977.
30. Lund EJ: "Bioelectric Fields and Growth." Austin: University of Texas, 1947, p 4.
31. Lund EJ: "Bioelectric Fields and Growth." Austin: University of Texas, 1947, p 24.
32. Brawley SH, Wetherell DF, Robinson KR: Electrical polarity in embryos of wild carrot preceeds cotyledon differentiation. Proc Natl Acad Sci USA 81:6064, 1984.
33. Overall RL, Wernicke W: Steady ionic currents around haploid embryos formed from tobacco pollen in culture. In Nuccitelli R (ed): Ionic Currents in Development." New York: Alan R. Liss, Inc., 1986, p 139.
34. Gamborg OL, Miller RA, Ojima K: Nutrient requirements of suspension cultures of soybean root cells. Exp Cell Res 50:151, 1968.
35. Rathore KS, Goldsworthy A: Electrical control of shoot regeneration in plant tissue cultures. Bio/Technology 3:1107, 1985.
36. Rathore KS, Goldsworthy A: Electrical control of growth in plant tissue cultures. Bio/Technology 3:253, 1985.
37. Goldsworthy A, Rathore KS: The electrical control of growth in plant tissue cultures: The polar transport of auxin. J Exp Bot 36:1134, 1985.
38. Niedergang-Kamien E, Skoog F: Studies on polarity and auxin transport in plants. 1. modification of polarity and auxin transport by tri-iodobenzoic acid. Physiol Plant 9:60, 1956.
39. Webster WW, Schrank AR: Electrical induction of lateral transport of 3-indoleacetic acid in the *Avena* coleoptile. Arch Biochem Biophys 47:107, 1953.
40. Hussey G: In vitro propagation of horticultural and agricultural crops. In Mantell SH, Smith H (eds): "Plant Biotechnology." Cambridge: Cambridge University, 1983, p 111.

Biotechnology in Agriculture, pages 53–81
© 1988 Alan R. Liss, Inc.

Somatic Embryogenesis and Polyembryogenesis in Conifers

D.J. Durzan and Pramod K. Gupta

Department of Environmental Horticulture, University of California, Davis, California 95616

————◆◆————

————◆◆————

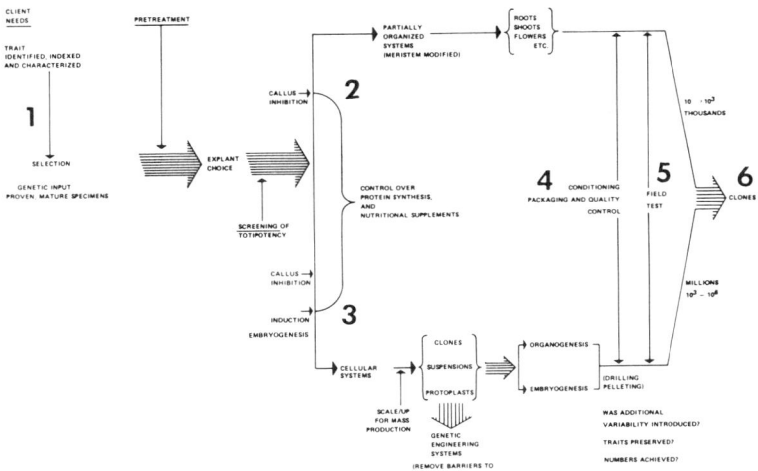

Fig. 1. Alternatives (items 2 and 3) in clonal production of elite conifer germplasm by cell and tissue culture as an aid to tree improvement. After recognizing client needs and identification of traits to be selected for (item 1) the donor may be pretreated and explants selected for micropropagation (item 2) or for somatic embryogenesis via SPE (item 3). The products arising from these alternatives require singulation, conditioning, packaging, and quality control (item 4) before field testing of products (clones, item 6).

Somatic polyembryogenesis (SPE) is a new physiological process that brings products of reproductive development into the laboratory for the capturing of genetic gains through tree-improvement programs. The gains captured by SPE are the result of additive and nonadditive genetic interactions (Fig. 1). Eventually SPE may be automated for large-scale production. For this to be achieved, we will need improved process controls for biological materials with highly variable physical properties. The machine-design background remains to be developed. Nevertheless, economies are projected not only for mass propagation of trees, but in terms of energy savings as well, because much of SPE occurs in darkness. With current micropropagation technologies, electricity and heat are major costs.

For the domestication and improvement of softwood trees, SPE provides a novel way to recover all products of a genetic cross, mass clone individual trees, create new genetic variation with somatic cells, and to reduce risks in germplasm availability by the continual provision and storage of somatic embryos as artificial seeds. SPE complements seed orchard technology where few alternatives existed before. Should SPE be applied to the mother tree by selection of explants from the haploid, gametophytic generation or from elite, proven, and mature specimens, SPE could significantly extend the value of current breeding practices. SPE has the potential for cost-effective mass propagation,[1] mechanization, artificial seed storage, preconditioning for precise environmental adaptation, and testing of genotype × environmental interactions.

SPE also provides a source of totipotent cells for studies with protoplasts for the direct introduction of new genetic information, especially where barriers now exist to sexual crossing. While considerable genetic variation exists for some species (Douglas fir, loblolly pine, Norway spruce), others, such as red pine, and some tropical *Pinus* species, have limited useful variation for breeding and tree-improvement purposes. SPE in suspension culture (Figs. 2–4,6) facilitates the search for useful new genetic variation, e.g., somatic aberrations, phenoclonal variations, and mutations (Fig. 4).

Our article provides some of the historical background leading to the claims in the introduction. It defines steps in the control of SPE and outlines prospects for new biotechnologies that may emerge from recent progress.

I. BACKGROUND

A. Historical Aspects

SPE was a natural outcome of principles of totipotency and somatic embryogenesis established by F.C. Steward's laboratory [36] especially with cell suspensions of the carrot (*Daucus carota*). Using the same technology, Durzan and Steward [15,16] attempted to generate somatic embryos from callus with cell suspension cultures of white spruce and jack pine zygotic embryos. For the next decade, efforts with conifers were directed toward evaluating a wide range of explant sources for their potential for somatic embryogenesis [10,12,13]. While embryolike structures were often observed, most of structures developed with internal cambial-like meristems that tended to show precocious vascular development [e.g., 12,13].

Embryolike structures (embryoids) were the focus of numerous reports (Table IV) and one patent [1]. The ultimate developmental fate of embryoids was not obvious until it was shown that further development proceeded by organogenesis and did not yield embryos with cotyledons [13]. The term

---

[1]Cost effectiveness currently relates mainly to high-valued horticultural varieties and to the provision of germplasm during poor seed years as in Douglas fir. For loblolly pine, genetically improved seed is usually available at costs which make most approaches to vegetative propagation (cuttings) less competitive. Cell and tissue culture methods are still far too expensive unless propagules are significantly genetically improved and proven superior in field tests.

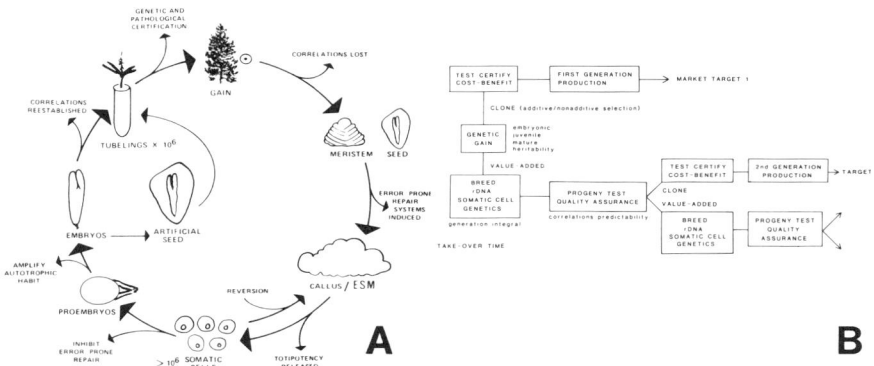

Fig. 2. Genetic gain in a cellular cloning cycle and production sequences for clonal products embodying the genetic gains. **A:** Genetic gains in conifers are captured by a cellular cloning cycle. At the top, the genetic gain is identified and explants (e.g., haploid or diploid from immature seeds) taken for the establishment of suspension cultures of somatic cells from callus or embryonal-suspensor masses (ESMs). Once suspension cultures are established, cells are transformed into somatic embryos as part of a mass cloning procedure. Alternatives include regeneration of plants through organogenesis and the development of hybrid plants from protoplasts rather than cells. **B:** Marketing of cloned ''genetic gain'' as somatic embryos or as artificial seeds. In forestry, our goal is to extend SPE to cover heritable gains in progeny-tested, proven, mature trees. After testing and certification, the first-generation clonal materials are introduced to meet a test-market target. Acceptance of clonal materials in a breeding program sets the stage for introducing new useful genetic variation by recombinant DNA (rDNA) and by somatic cell genetics to create a second-generation improved product.

''sphaeroblasts'' was introduced to distinguish these structures with cambial-like meristems from the hypothetical somatic embryos.

By 1984, other woody perennials (e.g., *Citrus* sp.) could be regenerated by somatic embryogenesis [35]. Krogstrup's thesis work [26] with *Picea abies* revealed embryolike structures in a white, slimy callus derived from epidermal cells of cotyledons and hypocotyls of germinating embryos. Soon after, Hackman et al. [24] reported somatic embryogenesis from immature seeds of Norway spruce. For other gymnosperms, Nagmani and Bonga [30] reported somatic embryogenesis from the haploid female gametophyte immature seeds of *Larix*. These embryos tend to be smaller, and their postulated haploid nature awaits confirmation.

During 1984–1985, Peter Krogstrup was repeating his observations at U.C. Davis. A project was initiated by P. Krogstrup, P.K. Gupta, and D.J.

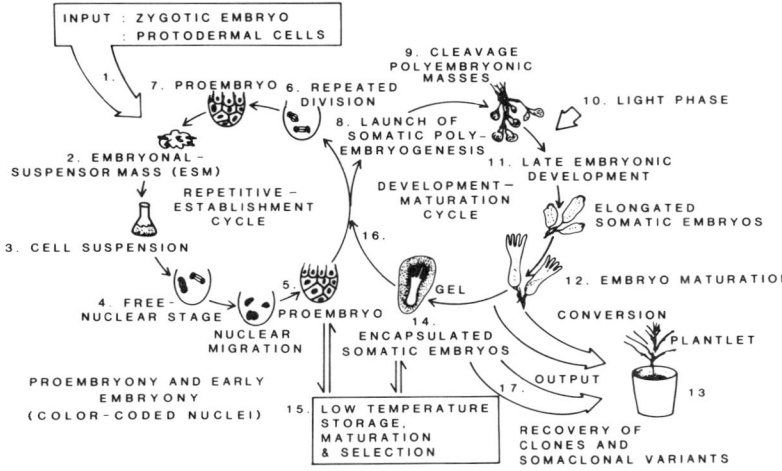

Fig. 3. Steps in a two-cycle process for somatic embryogenesis and polyembryogenesis. Totipotent cells are selected and introduced in an establishment cycle that is continually repeated to maintain embryonal-suspensor masses for the launch of embryogenesis. Once embryogenesis is launched in a development-maturation cycle, clonal products are drawn off for a variety of end uses, one of which is conversion to plantlets for clonal forestry. Stages 1 to 17 are described in the text.

Durzan to improve and extend earlier observations to other conifers and mature seeds, and to develop methods for cell suspension cultures. The outcome of this work led us to distinguish somatic embryogenesis from SPE.

B. Somatic Polyembryogenesis

In woody perennials, somatic embryogenesis has, historically and in practice, referred to the recovery of plants that represent the maternal genome, e.g., nucellar embryogenesis in *Citrus*. Somatic embryogenesis has been observed with gymnosperms [e.g., 18,20,21,24,26,31]; but usually, it is the offspring generation that is cloned. Somatic embryogenesis describes the clonal multiplication of genomes through apomictic processes that approximate zygotic embryogenesis and often involve an undefined ''callus'' which we try to avoid (Fig. 1.). By contrast, SPE was not readily apparent until our work with sugar pine [20], where zygotic, multiple proembryos from controlled crosses could be rescued from 5-year-old seeds. We found that the natural, zygotic, polyembryonic process could be repeated not with

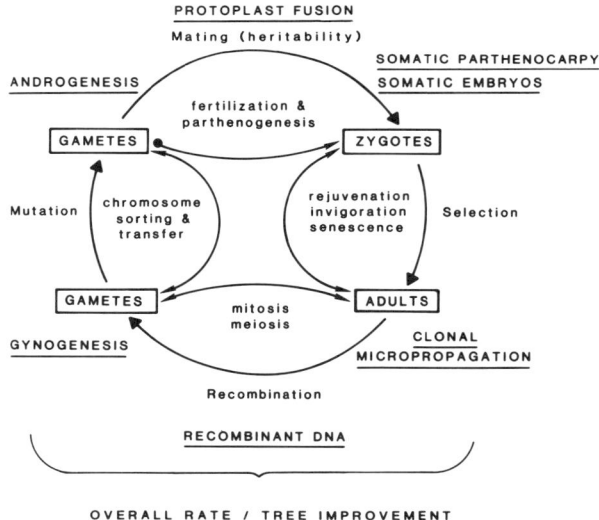

PROTOPLAST FUSION

Mating (heritability)

SOMATIC PARTHENOCARPY

ANDROGENESIS

SOMATIC EMBRYOS

fertilization &
parthenogenesis

GAMETES

ZYGOTES

Mutation

chromosome
sorting &
transfer

rejuvenation
invigoration
senescence

Selection

GAMETES

mitosis
meiosis

ADULTS

GYNOGENESIS

CLONAL
MICROPROPAGATION

Recombination

RECOMBINANT DNA

OVERALL RATE / TREE IMPROVEMENT

Fig. 4. Flowchart of fundamental stages in the life cycle of a woody perennial and processes used in classical genetics and in biotechnology for tree improvement. In *classical genetics* breeding strategies are based on mutations, selection, breeding (mating), and inbreeding (backcross breeding) that involve recombination of genetic information. In *biotechnology,* classical strategies can be extended to include somatic embryogenesis and parthenocarpy, clonal micropropagation of mature trees, and gametic apomixis (gynogenesis and androgenesis). Somatic embryogenesis should be extended, if possible, to include explants from adult trees and from reproductive cells (gametes). Clonal technologies differ from classical breeding methods because both additive and non-additive gains (e.g., position effects) are captured with morphogenic cells, whereas in sexual crosses only additive gains can be reliably sought. New genetic variation may be introduced into tree improvement by protoplast fusion, recombinant DNA methods, recovery of double haploids, microinjection, and electroporation of DNA.

callus, but with a proliferating embryonal-suspensor mass (ESM). The ESM is distinct from callus because products of division do not yield more callus. No transformation of cells in the ESM is needed to obtain embryos. Using ESMs, we have developed a process that has been process controlled to maintain embryo formation as long as the environment of the zygotic could be maintained. Indeed, a detailed analysis, confirmed with three other conifers, revealed that—

i. Daughter cells in the ESM repetitively produced proembryos with their suspensors in cleavage and noncleavage species.

ii. The predicted free-nuclear stages of an embryogenic "basal plan," which characterizes zygotic embryogenesis, could be found in cellular products of the ESM.

iii. A true-to-type polyembryonic developmental process, derived from the ESM, could be traced starting with free nuclei by a double-staining diagnostic test. Red and blue stains are specific for embryonal and suspensor cells, respectively, in the ESM and in zygotic embryogenesis [20,21]. Nuclei in cells of the ESM have differing developmental fates based on their properties and affinity for acetocarmine and Evan's blue.

iv. Embryo-specific acetocarmine-reactive mucilaginous products are produced by embryonal cells. These proteinaceous products are considered to represent components of the neocytoplasm in zygotic embryos according to the notion developed by Camefort [4]. In addition, acetocarmine is used in cytogenetics to detect chromatin. The term "relic cytoplasm" distinguishes the cytoplasm of cells that do not contribute to the new generation [34]. By contrast, suspensor cells are permeability to Evan's blue and may originate from the embryonal cells [19,22].

v. The fate of each cell type in the ESM could be traced after exposure to liquid nitrogen [19]. This exposure leads to the recovery only of viable, thawed embryonic cells that yield proembryos. From these surviving acetocarmine-reactive embryonal cells, suspensors are developed with their typical affinity for Evan's blue; i.e., the origin of the recovered polyembryonic cell masses was traced to red-staining embryonal nuclei and their associated neocytoplasm.

vi. In cell suspension culture, the highly differentiated ESMs often become polyembryogenic. In Douglas fir, which does not show polyembryogenesis in situ [2], polyembryogenesis can be induced in vitro, e.g., on semisolid surfaces and in cell suspension culture [42]. In suspension culture of loblolly pine, polyembryonic structures become lignified and form an infrastructure whereupon multiple embryos developed repetitively. The development of individual embryos and inhibition of polyembryogenesis can be controlled by abscisic acid (Fig. 5).

vii. The ESM retains the embryogenic potential, albeit severely reduced, even in mature seeds stored for up to 5 yr. ESMs recovered from immature seeds have been maintained for over 3 yr by 7–10-day subculture.

viii. SPE can be extended to other somatic cells of protodermal origin that contribute to the so-called somatic embryogenesis in "callus" [e.g., 24,26]. These morphogenic cells have been found in young needles from rejuvenated micropropagated axillary bud shoots from 60–100-yr-old Douglas fir [43]. The double-staining test [21] applied to these cells facilitates the search for other potentially embryogenic cells (ESMs) in tissues from all stages of the life cycle.

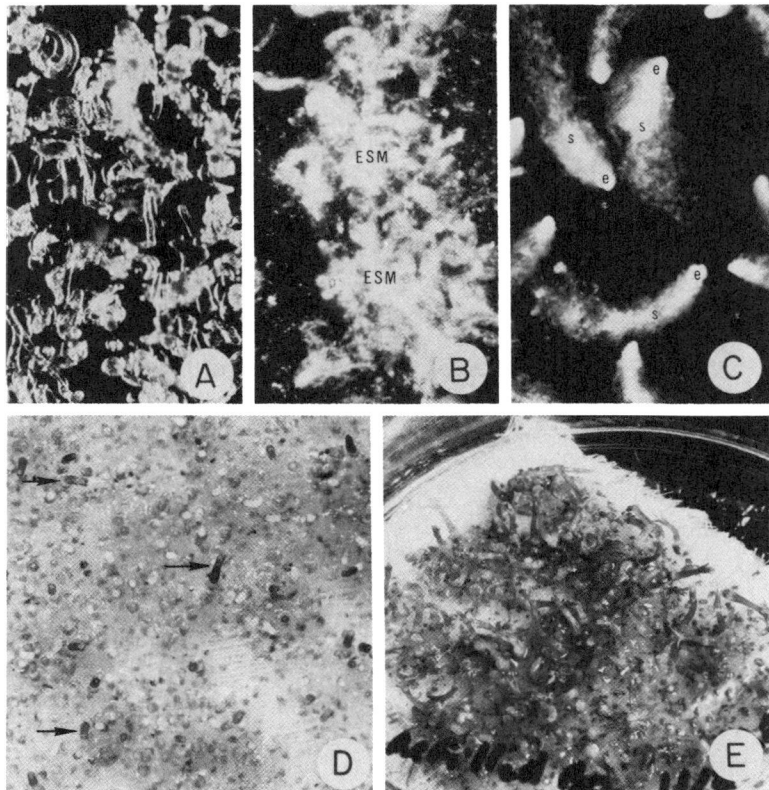

Fig. 5. **A:** Suspension culture of elongated (suspensor) and compact (embryonal) cells derived from an embryonal-suspensor mass (ESM) of loblolly pine. Elongated cells average about 100 μm in length. **B:** Before the launch of somatic polyembryogenesis the developing proembryos of loblolly pine develop a fabric of ESM that should be avoided as in Figure 5C. × 40. **C:** Addition of abscisic acid inhibits the polyembryogenic fabric and enables individual embryos (e) to develop with their suspensors (s). × 38. **D:** When Norway spruce somatic embryos are plated onto a cheesecloth support in liquid medium, individual plantlets (arrows) begin to sprout from the surface. **E:** Plantlets of Norway spruce obtained by polyembryogenesis from cell suspension cultures plated onto cheesecloth. Plantlets are at various stages of elongation. Those with cotyledons vary from 0.5 to 1.0 cm in length.

Fig. 6. **A:** Alginate-encapsulated somatic embryos of Norway spruce, ca. 5 to 8 mm dia. Singulation of encapsulated embryos facilitates storage and machine handling. **B:** Germination of an encapsulated somatic embryo of Norway spruce ca. 2 cm long. **C:** Douglas fir plantlet (2 cm high after 3 wk) growing in soil and derived by somatic embryogenesis in cell suspension cultures. **D:** Loblolly pine plantlet (8 cm after 6 wk) growing in soil and derived from cell suspension cultures of embryonal-suspensor masses.

ix. From factors contributing to SPE, a model based on metabolic networks, physiological states, and signal theory could be developed [cf. 10,14]. This work is aimed at increasing our analytical prowess so as to develop model-reactive adaptive control systems that may be appropriate for the scale-up and mechanization of SPE [44].

C. Model Processes and Strategies

Attributes and stages of zygotic embryogenesis provide the rationale and criteria (model) for SPE [cf. 4,8,31,34]. Embryo rescue from the ESM was fundamental in providing criteria for improved process control of somatic cells in vitro. SPE had its origin from studies attempting to induce somatic embryogenesis on agar plates with the ESM and then with cell suspension cultures (Table IV). It was, however, work with cell suspensions that enabled i) the separation of cells in the ESM for improved microscopic observation and their cytochemical distinction from callus, ii) study of physical and chemical factors in "process control" in SPE, and iii) recognition of new requirements and aberrations in embryogenic development, such as the need for abscisic acid (ABA) to control individual embryo development and avoid the production of sphaeroblasts.

Also, because of the complexity of zygotic embryogenesis in conifers, baseline information is required to define a) natural variation in zygotic embryogenesis; b) how zygotic embryos grow and develop outside of the ovule; c) nutrient media, plant growth regulators, and conditions that mimic and accurately represent the internal environment of the developing seed; d) suitable explant sources; e) indicators for screening cells for their potential for somatic embryogenesis; f) methods that reduce the handling of cells and developing embryos; g) methods that predict the developmental outcome of cells in relation to experimental variables; h) methods to singulate and preserve somatic embryos for long-term storage; and i) conditions for embryo conversion to plantlets in soil.

Ideal characteristics of model systems (Tables I, II) should relate to current strategies for capturing genetic gains [38,39]. For conifers, components of the genetic gain involve the following:

1. Heritability. The proportion of the total phenotypic variance for which genetic differences are responsible.

2. Selection differential. The average superiority or mean phenotypic value of the individuals selected as parents, expressed as a deviation of the mean phenotypic value of the population of their origin. Selection is used to choose individuals or populations with desirable traits for genetic improvement.

3. Embryonic/juvenile/mature correlations. Coefficients that measure

TABLE I. Ideal System Attributes for Developing Biotechnologies With Somatic Cells of Conifers

A. Characteristics of ideal system[a]
  1. Uniform population of cells or cell types
  2. Ability to control the isolation of factors and predetermine and maintain the critical variables largely by physical means
  3. Sufficient levels of chemical factor(s) for purification, characterization, and application
  4. Controllable fluctuations in critical components that are affected by nutritional and environmental variables.
  5. Capacity for baseline monitoring and evaluation in process control
  6. Availability of parental trees or mutants for studying the origin and molecular specificity of the trait or gain sought
  7. Availability of well-defined nutrients or simple methods for perturbing the critical factor(s)
  8. Ease of study by physical and chemical means
  9. Capability for demonstrating relations between results in the culture system and events in nature or under field conditions
B. Advantages of cloning from cell suspensions[b]
  1. Few cells as starting material
  2. Mother tree can be used for other purposes
  3. Cells grown aseptically and are totipotent (i.e., regenerate whole plants and life cycle)
  4. Space and system of propagation is economical and controlled
  5. Suitability for applying microbiological and molecular genetic methodologies and ease of scale-up in bioreactors
  6. Mutations can be introduced and identified using large numbers of somatic cells
  7. Populations of cells can be rapidly screened and selected for their elite traits provided that efficient procedures and baseline knowledge are available
  8. Efficiency and time for recovery of plants (cell to nursery 10–12 mo)
  9. Efficiency of inserting new traits is increased; however, the list of useful traits to be inserted into cells or protoplasts is not yet large
  10. With haploids, mutant traits are not masked as in diploids; diploidization of haploid cells to produce homozygotes allows for improved genetic analysis and widens opportunities in tree breeding

[a]Characteristics useful in molecular genetic studies.
[b]See Figure 2.

the degree of association of given traits between two or more stages of the life cycle. For Douglas fir, up to 9 yr of evaluation are required for assurance that trees will perform as expected when maturity or harvest dates are reached.

4. Generation integral. Time and resources may be saved or inadvertently lost by application of current biotechnologies based on the long life cycle of trees. For each generation, we need to search for methods and systems that efficiently reduce the length of component processes in the life

TABLE II. Comparison of Properties of Two Complementary Breeding Systems Aimed at Propagating Trees*

Factor	Tree	Cell
Growth cycle	6 or more yr	25 to 26 hours (cell cycle)
Size	Up to 250 ft high	50–100 μM diameter
Space (10^6 progeny per year)	2.5–10 acres of seed orchard	10 liters of suspension culture
Numbers to		
1) detect mutation	10^7 trees	10^7 cells
2) detect linked traits	10^3 trees	10^3 cells
Time to produce bulk seed or somatic embryo	8–13 yr, minimum of 2 to 3 yr for cones on seedlings (9.7 $\times\ 10^6$ trees from 25 orchards over 16 yr)	1 to 2 yr estimated (10^6 per month) (somatic embryos)
Seed production predictability	Variable seed years (species dependent)	Controlled production in vitro on demand
Uniformity of genetics	Variable, selection required, controlled pollination	Homogeneous, low error frequency, easy rogueing, also somaclonal aberration and variation easily induced
Propagation	Vegetative (cuttings), grafting, rooting (sometimes difficult)	Up to 500 copies per explant by organogenesis (tissues, organs), millions by embryogenesis (research needed to establish this alternative)
Ploidy	Haploids difficult to produce	Some haploids easily cultured
Flexibility in breeding system	Conventional barriers to crossability	Apply methods of microbial and molecular genetics; barriers to crossability can be partially removed; genetic engineering can be attempted
Types of breeding system Mutation Inbreeding Hybridization Backcross Selection	Selection most applicable; hybrids possible for certain species; other methods slow and require 8–10 yr generation times	Most breeding systems can be exploited with cells and their protoplasts[a]; field testing required for certification of quality and "true to type"
Cost to maintain system	$1,500/year/hectare (1979)	$1,500 per liter (estimated)
Nos. required (estimates)	570 × 10^{-6}/year (Quebec) 4 to 7 × 10^9/year (Ontario) 100 × 10^6/year (British Columbia) 1 × 10^9/year (USA)	Not established

*The first system involves the need for a seed orchard and the other a viable cellular cloning cycle [after reference 12].

[a]With haploid tissues in a female cone we may be able to recover products of meiosis by somatic embryogenesis and parthenogenesis [44].

cycle, in the rotation cycle and in seed orchards (breeding, production, and testing periods).

In summary, capturing genetic gains involves heritability, selection, correlations, and the generation integral [39].

II. PROCESS SEQUENCES (SPE)[2]

Sequences for the production of somatic embryos in suspension culture are illustrated in Figure 3. Several stages in cell suspension culture are shown in Figure 5.

A. General Sequence

1. Surface sterilize cones, seeds, shoots or roots [18,20,21] (input, Fig. 3).

2. Excise ovules with female gametophyte and remove embryos with suspensor mass or remove explants bearing protodermal cells from donor.

3. Isolate the embryonal-suspensor mass (ESM) by maintaining contact of ESM with original explant and culture medium in darkness. ESM is characterized as white, slimy, proliferating morphogenic mass of cells emerging from the zygotic proembryo or from protodermal cells of shoots and roots. ESM is distinct from callus (nonembryogenic by definition unless transformed) and its proliferation is evoked by high 2,4-D (50 μM) (item 2, Fig. 3). Zygotic embryos may be removed and separated from ESM for rescue (optional).

4. For explants with protodermal cells, induce ESMs with 2,4-D or a combination of plant growth regulators [e.g., 24,26,43].

5. Proliferate the ESM with or without zygotic embryos by subculture on agar with low 2,4-D (ca. 5 μM) every 9 to 10 days. ESM can be transferred to the liquid medium of the same formulation to establish cell suspension cultures in darkness. Adjustment of subculture rate, inoculation density, and aeration (ca. 10 g per 100 ml and 1–50 rpm shake rate) are required (item 3, Fig. 3). Excessive rotation or centrifugation are inhibitory to embryonic development.

6. Select ESM types by light microscopic inspection and double-staining

---

[2]The following abbreviations are used in this review: ABA = abscisic acid; BAP = N^6-benzyladenine; BM = basal medium; CH = casein hydrolysate (Difco Casamino Acids); 2,4-D = 2,4-dichlorophenoxyacetic acid; gln = L-glutamine; KN = kinetin; MS = Murashige-Skoog medium (1962).

methods; i.e., provide evidence for stages "conifer-type" somatic embryogenesis and polyembryogenesis (items 4, 5, Fig. 3).

7. Prospective somatic proembryos arise as a "basal plan" from primary embryonal cells in the ESM. Embryonal cells are maintained by mitosis in low 2,4-D and under these conditions often reveal a free-nuclear stage. Usually a characteristic number of nuclei are produced (e.g., four in loblolly pine) that stain differentially with acetocarmine or Feulgen and Evan's blue (items 6, 7, Fig. 3).

8. The nuclei associated with neocytoplasm migrate as in the "basal plan" for conifer-type proembryogenesis [34]. Nuclei that contribute to the formation of the proembryo stain red with acetocarmine and Feulgen. Other nuclei, which stain more intensely blue with Evan's blue, usually contribute to the formation of the suspensor (items 4, 6, Fig. 3). Weakly sustaining cells may represent callus that eventually turns brown especially if cultural conditions required for ESM production are met. We have found it very difficult to maintain nonembryogenic calli under conditions that promote somatic embryogenesis and polyembryogenesis.

9. Within the ESM, some suspensor and proembryonal cells contain red-staining nuclei. These are believed to be the source of the repetitive mitotic process that contributes to the formation of polyembryonic clusters normally produced by cleavage as in the zygotic seed (cycle 1, Fig. 3).

10. As multiple embryos develop within the ESM, the primary embryonal cells proceed through true-to-type proembryonal stages of development. Nuclei (often staining blue) contribute to the formation of the suspensor in a pattern approximating that described as the primary upper tier [cf. 34] (cycle 1, and item 8 in cycle 2, Fig. 3).

11. The ESM may be maintained productively for several years in darkness by subculture (cycle 1). This process is referred to as the repetitive establishment cycle. The quality of cells and potential for proembryonal mass development can be monitored by single or double diagnostic staining methods (cycle 1).

12. Upon demand and at will, early embryonic development is launched by transfer of cells as an ESM or in cell suspension to subsequent media (see Culture Conditions). Media need to be adjusted for appropriate levels of plant growth regulators and nutrients. The ESM is maintained in darkness till the late globular state of development. Subculture rate is maintained at approximate 1–2-wk intervals (item 8, cycle 2, Fig. 3).

13. When the globular embryonic masses (ca. 0.5–1.0 mm embryo diameter) are obtained, their suspensors are already elongated. In some cell suspensions, these masses show lignification of suspensors in polyembryonic clusters. Lignification imposes some rigidity on the ESM. Rigidity facilitates

subculture, especially in cell suspensions. If such a mass develops by prolonged suspension culture, a fabric of embryos is formed that may be difficult to separate. Treatment with ABA (1×10^{-6} M) inhibits the formation of polyembryogenic masses and allows growth of individual embryos. Globular embryos retain their affinity for red and blue stains (item 9, Fig. 3).

14. The globular embryonic masses are launched into late embryogeny by transfer to the appropriate culture medium designated for each species (see Culture Conditions). Cultures are then grown in the presence of diffused white light (e.g., 5.0, 2.0, and 0.5 μW cm^{-2} in the red, blue, and far-red spectrum) (item 10, Fig. 3).

15. Each embryo, still bearing a small ESM, completes its development in 3–4 wk. Embryos develop a variable number of cotyledons, as they do in nature (items 11, 12, Fig. 3). Embryos may be singulated by several means; e.g., embryos can be encapsulated at this or at an earlier globular stage [U.S. Patent No. 4,217,730] for storage at 4°C or in liquid nitrogen or until further use. Encapsulated embryos can be converted to, or germinated as, plantlets, or fed back into the establishment cycle for cloning. The latter is important if potentially useful somaclonal aberrations arise that require scale-up (items 14, 15, 16, Fig. 3).

16. Converted embryos are planted in soil and grow into plantlets that represent the new generation (item 17, Fig. 3).

17. Initial process control is achieved by fine-tuning culture parameters for each ESM and by adjusting levels of 2,4-D, auxins, cytokinins, abscisic acid, *myo*-inositol, and nitrogenous supplements that contribute to conditions fostering embryo rescue.

B. Culture Conditions

The following general protocol can be followed for loblolly pine [21], Norway spruce [18], sugar pine [20], and Douglas fir [42]. Further fine-tuning for these and other conifers having precise genetic adaptation may be required [33]. Initially, tissues and ESMs are inoculated every week on culture media (cf. Table IV).

We use an MS basal medium [20,21] with modification of NH$_4$NO$_3$ (550 mg/liter), KNO$_3$ (4,674 mg/liter), and thiamine · HCl (1.0 m/liter). The following media are used for species listed in parentheses. Half-strength modified MS (*Picea*), DCR (Douglas fir), and LP (loblolly pine) media [cf. 27] are supplemented with *myo*-inositol (1,000 mg/liter), sucrose (3%), L-glutamine (gln) (450 mg/liter), casein hydrolysate (CH) (9,500 mg/liter), 2,4-D (5×10^{-5} M), kinetin (KN), and N^6-benzyladenine (BAP) (each at 2×10^{-5} M) at an initial pH of 5.7 before autoclaving. Cultures are maintained on 0.6% agar (Difco Bacto agar) plates in darkness at 23°C ± 2°C.

To complete early embryony, the proliferating ESM is subcultured to the same medium as described above except for 2,4-D (5 × 10^{-6} M), KN, and BAP (each at 2 × 10^{-6} M). After four or five subculture, ESMs are transferred to lower concentrations of 2,4-D (1 × 10^{-6} M) and KN, BAP (0.25 × 10^{-6} M) with ABA (1 × 10^{-6} M). After three or four subcultures, the globular stage of embryogenesis is fully evident. Embryos elongate and develop cotyledons within 8–10 wk at 25°C ± 2°C when transferred to a sterile-filtered liquid medium, with filter paper support, without growth regulators and under continuous white light (5.0, 2.0, 0.5 μW cm² nm^{-1} in blue, red, and far-red, respectively) (Fig. 4). Complete plants are developed in a basal medium containing 0.25% w/v activated charcoal (Merck, NJ), and sucrose (1%). Casein hydrolysate and glutamine are removed. This subculture sequence completes the recovery of plantlets from callus on agar plates.

For cell-suspension culture, embryonal-suspensor masses (ca. 2 g in 50 ml) are placed in shaking (120 rpm) 250-ml Erlenmeyer flasks with fluted bases. The culture medium is 0.5 strength basal medium (see above) with 2,4-D (5 × 10^{-6} M), KN, and BAP (each 2 × 10^{-6} M). Cell suspensions are formed rapidly in darkness when subcultured every 5–6 days. Repeated subculture produces well-dispersed suspensions of single cells, aggregates of two to five cells, and larger embryonal-suspensor masses. Packed-cell volume is measured after centrifugation of cell suspension of each flask at not greater than 250g for 10 min.

Currently the best recovery of plantlets has been obtained with Norway spruce. Transfer of cell suspensions to cheese cloth supports has yielded 225 ± 25 plantlets from 5 ml of cell suspension (M. Boulay, P.K. Gupta, P. Krogstrup, and D.J. Durzan, Plant Cell Reports 1988, in press).

C. Histological and Cytochemical Staining

Cells in suspension cultures or in callus are stained as follows. Samples of cells, (packed-cell volume of 5–10 μl) are suspended in liquid medium to which 2% acetocarmine (1:1 v/v) is added. Cells are heated slighted for 15 s and filtered to remove excess stain. Evan's blue (0.5%, 1:1 v/v) is added to the acetocarmine-stained cell suspension and washed with medium and filtered to remove excess of stain. After double-staining, cultures are resuspended in 100% glycerol to improve optical clarity of cells on slides for microscopic inspection and the distribution of dyes followed microphotographically. The process can be repeated with Feulgen and Evan's blue [21].

D. Encapsulation and Freeze Preservation of Somatic Embryos

Somatic embryos are dipped in 1% w/v sodium alginate (Protan, Norway) coming from a separatory funnel. After exposure to alginate, the coated embryos are dropped in a beaker containing 100 mM calcium nitrate and stirred for 8–10 min [21]. Encapsulated embryos are then washed with sterile water to remove excess calcium nitrate. Embryonal-suspensor masses are freeze-preserved, and embryos recovered from thawed cells as described in references 19 and 40.

III. APPLICATIONS

With over a century of high-grading of trees [e.g., 29] and with the increasing demands on existing forests on a global scale, efficient methods for mass cloning in tree-improvement programs are required [11,12]. The demand for improved trees can be conceived as being equivalent to the genetic gain embodied these in trees (Fig. 2B). New methods are continually being sought to remove barriers to productivity and breeding based on insects, diseases, incompatibility, natural disasters, impact of man (acid deposition), etc. Biotechnology brings the forest conveniently into the laboratory at critical stages of the long, complex life cycle of coniferous trees.

Somatic embryogenesis impacts on horticulture and on mass-production forestry by enabling the clonal mass production of elite trees, germplasm enhancement through selection and breeding strategies that introduce useful genetic information, and risk reduction in production forestry through the development of early diagnostic methods and by certification of germplasm.

A. Artificial Seeds

Somatic embryos can be separated and coated with alginate or other materials for handling and storage (see Chapter 9, this volume). Someday artificial seeds may become a reality. These would have attributes based on a) introduction of agricultural chemicals, fungicides, microorganisms, etc., into coatings; b) convenience of precision planting; c) seedling uniformity, synchrony, and yield; d) possibly lower costs (for high-valued specimens) vs. genetically improved seed; e) efficiency of year-round production processes leading to the marketplace; f) independence of problems of poor seed years, recovery of seeds from tall trees, and competition for seed with animals and insects.

While we are a long way off from producing artificial seeds (cf. Fig. 6A), a supply of elite germplasm in the form of somatic embryos is a prerequisite to developing this technology. Much more needs to be learned about the biology of seeds, e.g., compartmentation, compatibility, quality, and slow release of nutrient reserves that could constitute an artificial female gametophyte.

B. Germplasm Storage and Preservation

We can anticipate that somatic embryogenesis and polyembryogenesis will provide a wide spectrum of clonal materials from true-to-type to highly aberrant [44]. Years of field testing will have to pass before the utility of this germplasm can be demonstrated. For this reason, cryopreservation of ESMs in liquid nitrogen or low-temperature storage of encapsulated embryos may provide a convenient storage medium until the appropriate tests are performed. Moreover, we suspect that the cryopreservation process itself imposes usefully selective parameters that remove potentially weak individual trees. Thawed cells have demonstrated a survival which could be of use in the search for cold-tolerant properties. We have already recovered Norway spruce and loblolly pine embryos from embryonal cells exposed to liquid nitrogen [19]. More work is needed to develop definitive processes and tests in this subject-matter area. Cryopreservation of totipotent cells may someday complement existing strategies for the preservation of germplasm resources that can be maintained "on the stump" in protected forests.

C. Germplasm Enhancement

Where existing useful genetic variation is limited, biotechnologies based on recombinant DNA technology, protoplast fusion, microinjection, electroporation, etc., may soon provide useful new opportunities for tree improvement [9,17,22,23,41]. Direct gene insertion could help overcome some problems associated with incompatibilities typical of interspecific genetic crosses. In unpublished work (Gupta, Dandekar, Durzan) we have introduced into protoplasts by electroporation the luciferase (*luc*) gene from fireflies into Douglas fir and loblolly pine. As cell walls are regenerated and luciferin is added to the culture medium, *luc* activity is expressed visually as protoplasts "light up." While this gene has little immediate commercial utility, it serves as a "reporter" gene to follow how efficiently the foreign gene is transferred to and expressed in daughter cells. The fate of this gene has been followed in tobacco plants by watering plants with a weak solution of luciferin.

Nevertheless, while there is now considerable evidence that plant cells can be genetically modified, we must recover commercially useful plants, products, and services from these approaches. Before this can be done convincingly, we need reliable plant regeneration methods for genetically modified cells and the appropriate field testing of products.

With the advent of SPE, we now have a source of totipotent protoplasts that could be genetically manipulated. Preliminary results in our laboratory with loblolly pine, Douglas fir, and Norway spruce have shown that the early stages of SPE can be recovered from protoplasts derived from embryonal cells in the ESM [22] (P.K. Gupta, M. Boulay, and D. Durzan, unpublished). With genetic engineering; protoplast hybridization, including haploid cell fusion or doubling and the recovery of somatic embryos, we can now bring the recombinant phases into the laboratory. Products of crosses and genetic manipulations can be attempted all year around and independently of the vagaries of meiosis, environment, pollination, and fertilization. We have yet to explore and evaluate the utility of pollen and the biotechnology that derives from the reproductive phase of the life cycle.

Recovery of somatic embryos by SPE from protoplasts opens the door for direct genetic modification; protoplast fusion; and hybridization, liposome integration, electroporation, reduction of lethal genetic loads by selection and rogueing, and technologies never before possible with forest trees. In the short term, we can anticipate significant improvements in current plant genetic engineering methods using *Agrobacterium tumefaciens* [6,17,41]. The use of *Agrobacterium* may be bypassed through the introduction of foreign genes by electroporation provided that the electroporated cells remain viable and morphogenic. This approach could speed up genetic engineering and obviate steps that involve removal and cleanup of cells cocultured with *Agrobacterium*. In the long term, the effectiveness of any method depends on how well foreign genes are tolerated, integrated, and expressed in a timely and true-to-type fashion throughout the life cycle. For this reason, we may still have to face problems related to error-prone repair mechanisms and incompatibilities associated with the genetic apparatus [e.g., 8,9] and its neocytoplasm during development.

D. Seed Orchards

A seed orchard is a plantation established primarily for the production of seed of proven genetic quality. A clonal seed orchard is composed of vegetatively propagated (usually grafted and compatible) trees established for the production of seed of proven genetic quality from the clones themselves.

Seed orchards are a considerable investment for any forestry enterprise. Efforts aim to maximize benefits and minimize costs of seed orchards. It is usually the lack of regular, heavy, flowering years that limits seed recovery for tree improvement programs [e.g., 7]. Furthermore, by the time trees reach commercial seed production, their height and competition with squirrels or insects may pose problems for the harvesting of cones, fruit, and seed [25]. Good crops are not obtained every year, nor do all genotypes contribute equally to pollen or seed production [37]. In some instances, dwarfing rootstocks and miniaturization of seed orchards have been proposed to deal with large, unruly, and unpredictable orchards [28].

Through biotechnology, and specifically with SPE, we have another set of options to consider, viz., the production of artificial seeds by SPE "at will and on demand" in a laboratory associated with seed orchards. In recent years, the biotechnologies which were required for the regeneration of coniferous species have materialized or at least have become close at hand (Table IV). Thus, one of the greatest opportunities is to revolutionize strategies associated with current practice of seed orchards. In North America, seed orchards are sometimes 25 acres and very costly and unpredictable to operate. Opportunities are compared in Tables II and III.

A small laboratory designed for artificial seed production and certification would be a great complement to current field-based seed orchards. We might anticipate that in the next decade much of the experimental approach genetic improvement of trees could be attempted in such laboratories. The benefit of this approach will relate initially to developing baseline genetic and physiological information.

Limitations in somatic embryogenesis are based on the totipotency of embryonal cells from all stages of the life cycle, how well these cells carry foreign genetic information, and how products utilizing artificial seeds or their counterparts can be introduced into current reforestation and afforestation programs. These are not trivial problems, and their solutions are worthy of investigation.

E. Biotechnology and the Life Cycle of Trees

The long life cycle and size of conifers lends credence to their precise genetic adaptation to a wide spectrum of microclimates. For SPE to become truly useful we should be able to evoke the process in explants from tissues representing most stages of life cycle (Fig. 4). Furthermore, methods should be sought not only to invigorate and rejuvenate cells for propagation, but also to accelerate maturity as has been done with *Prunus* cell suspensions [14]. The clonal production of elite trees and their somaclonal or phenoclonal aberrations should enable the testing of genotype × maturity state at locations across wide targeted geographical ranges. One nutritional param-

TABLE III. Comparison Between Seed Orchards (for Producing Improved Seed in Quantity) Versus Cell and Tissue Culture Methods (for the Mass Propagation of Elite Hybrids by Somatic Embryogenesis and Polyembryogenesis)*

Factor	Seed orchard	Cell and tissue culture
Selection of phenotype/genotype	Natural or planted forest	Natural or planted forest

Reestablishment for seed orchard	From seed provenances
	  Grafting (compatibility)
	  Rooting
		From live cells (haploid
		  diploid) displaying
		  totipotency

Ingredients and processes	20–30 clones (avoid inbreeding)	Not relevant
	Isolation to avoid pollination	Not relevant
	Selection of locality for orchard	Controlled laboratory
	Promotion of early flowering	Not relevant
	Avoid inbreeding by randomization	Not relevant
	Maximization of seed yield	Maximization of totipotency[a]
	Fertilization	Nutrient medium
	Wide spacing	Few liters per clone
	Bending	Not applicable
	Girdling	Not applicable
	Root pruning	Not applicable
	Irrigation	Not applicable
	Weed control	Process control
	Animal, insect and disease control	Aseptic conditions Artificial seed or tubeling
Requirements to produce 10^6 trees	Viable seeds 13/cone	$>10^6$ somatic embryos
	Cones/ramet 125	Not applicable
	Seedlings/ramet 1,625	Seedlings/cell 1[a]
	Plantable trees/ramet 541	Trees/cell 1[a]
	Ramets 1,848	Cells 10^8
	Ramets per acre 170 (16 × 16′)	Cells/liter 10^6
	Acres 10–12	Laboratory space 2,000 ft^2
Goals	Improved seeds in quantity	Mass propagated elite hybrids Storage of artificial seeds Introduce new variation (genetic)

*The goals of cell and tissue culture technology do not replace, but complement, the existing and traditional seed orchards for improved seed production [after reference 12]. Seed orchard information based on [5].
[a]Polyembryogenesis may extend these values substantially.

eter of interest is the efficiency by which trees use available nitrogen. For this work, we will need decades of more baseline nutritional, cytogenetic, physiological, and biochemical studies. Unless these are done, much of the potential of forest biotechnology will be constrained and confounded by ignorance.

TABLE IV. Examples of Somatic Embryogenesis in Conifer Species: Some Relevant Observations According to Species*

Species	Explants	Media	Response	References
Black spruce (*Picea mariana*)	Immature embryos	Modified basal medium + 2, 4-D + BAP	Somatic embryos	[T1]
Douglas fir (*Pseudotsuga menziesii*)	Cotyledons 1–4 yr old trees (meristemic) tissue	Modified MS + NAA + BAP Modified medium + amino acids + amide supplement + NAA + BAP	Embryoids Proembryolike structures with aberrant development sphaeroblasts	[T2] [T3, 4]
	Immature embryos and mature embryos	DCR medium + casein hydrolysate + gln + 2, 4-D, + KN + BAP	Somatic embryos	[T5]
Jack pine (*Pinus banksiana*)	Seedling explants (hypocotyl)	White's basal medium + NAA + coconut milk, Steinhart et al. (1961) KN & seed extract	Proembryolike structures Free nuclei in elongated cells single suspensor-like cells	[T6] [T7]
Loblolly pine (*Pinus taeda*)	Immature embryos	Modified MS and LP + casein hydrolysate + gln +2, 4-D + KN + BAP	Somatic embryos, plantlets, and encapsulation	[T8]
Norway spruce (*Picea abies*)	Immature embryos	Von Arnold and Eriksson basal medium + casein hydrolysate + 2, 4-D + BAP	Somatic embryos plantlets	[T9–11]
	Mature embryos	Von Arnold and Eriksson medium + 2, 4-D + BAP Modified MS + casein hydrolysate + gln +2, 4-D + KN + BAP	Somatic embryos Somatic embryos plantlets and encapsulation	[T12] [T13]

Sugar pine (*Pinus lambertiana*)	Cotyledons	Modified MS + casein hydrolysate + gln +2, 4-D + KN + BAP	Somatic embryos	[T14]
	Mature embryos	DCR medium + casein hydrolysate + gln +2, 4-D + KN + BAP	Somatic embryos plantlets	[T15, 16]
White spruce (*Picea glauca*)	Seedling explants (hypocotyl)	White's basal medium + coconut milk + NAA	Proembryolike structures	[T6]
	1–4 yr old trees (meristemic tissue)	Modified MS + amino acids (R-medium) + amide supplement + NAA + BAP	Embryoid sphaeroblasts	[T4]
	Immature embryos	Modified basal medium +2, 4-D + BAP	Somatic embryos	[T1]

*Abbreviated references

T1.   Hakman I, Fowke LC, Rennie PJ VI Int Cong Plant Tissue and Cell Culture, Minnesota, USA, 1986 p 194.
T2.   Abo El-Nil: U.S. Patent No. 4,217, 370, 1980.
T3.   Durzan DJ: "Plant Cell Cultures: Results and Perspectives." F. Sala et al. (eds). Amsterdam: Elsevier 1980, pp 283–288.
T4.   Durzan DJ: Proc 5th Int Cong Plant Tissue and Cell Culture, Tokyo, Japan, 1982, p 113–114.
T5.   Durzan DJ, Gupta PK: Plant Science 52:229–235, 1987.
T6.   Durzan DJ, Steward FC: Proc Int Union Forest Res Organization, Helsinki, Finland, 1970, p 1–18.
T7.   Durzan DJ, Bennett DR: Proc 11th Mtg Committee Forest Tree Breeding Canada, Environment Canada, 1968, pp 19–21.
T8.   Gupta PK, Durzan DJ: Biotechnology 5:147–151, 1987.
T9.   Hakman I, Fowke LC, Von Arnold S, Eriksson T: Plant Sci 38:53–59, 1985.
T10. Hakman I, Von Arnold S: J Plant Physiol 121:149–158, 1985.
T11. Becwar MR, Noland TL, Wann SR: Proc Int Congr Plant Tissue and Cell Culture, Minnesota, USA, 1986, p 137.
T12. Von Arnold S, Hakman I: J Plant Physiol 122:261–265, 1986.
T13. Gupta PK, Durzan DJ: *In vitro* Cell Dev Biol 22:685–688, 1986.
T14. Krogstrup P: Can J For Res 16:664–668, 1986 and PhD thesis, Royal Veterinary and Agric. Univ., Copenhagen, Denmark (1984).
T15. Gupta PK, Durzan DJ: Biotechnology, 4:643–645, 1986a.
T16. Gupta PK, Durzan DJ: In Proc VI Int Cong Plant Tissue and Cell Culture, Minnesota, USA, 1986. p 394.

F. Somatic and Reproductive Cell Genetics

The long and complex life cycle of conifers reduces the efficiency of many of the current plant breeding strategies. Because of our newly acquired ability to cultivate and manipulate somatic cells [3], the opportunity arises to develop the subjects of somatic and reproductive cell genetics. Especially useful is the ability to recover the products of meiosis and fertilization by parthenogenesis, cultivation of ESMs and by embryo rescue. If and when this is coupled with cytogenetics [32] and the recovery of embryos from cells of tissues at any stage of the life cycle, we will be in a better position to understand the role of genes in development and the ease by which *"sports"* or mutations and hybrids can be produced. Especially significant is the potential to explore directed totipotency with neocytoplasm to amplify secondary-product formation and to introduce synthetic DNA or genome-suppressed DNA (codons and anticodons) for gene expression studies. We may even be able to exploit chronobiology and speed up the evolution of useful conifers [12]. The key to success is better understanding of selection procedures for the quality control of genetically modified somatic cells, and some understanding of how these genetically engineered cells and trees might benefit humanity.

IV. PROSPECTS

Several opportunities and problems have been cited under Applications; however, to make realistic projections we should consider existing barriers to progress [12,23]. We need to understand better model zygotic and somatic embryonic systems for the development of model-reactive process controls [44]. Conversion of somatic embryos to soil, while currently low, has the potential of being increased dramatically, as judged by the large number of acetocarmine-reactive cells in suspension culture. SPE should be sought from other tissues, e.g., haploid female gametophyte, protodermal cells at shoot apices, gametes, etc. Indeed, with a rapid diagnostic test such as acetocarmine, we may be able to faciliate screening for totipotent cells from mature donor and elite proven trees.

Much of the embryo progeny testing may be done in the laboratory to establish better juvenile-mature correlations and gene × environment effects in the field. Clonally propagated germplasm will not wholly supplement seed except possibly in specific horticultural applications and markets.

SPE offers a new way of increasing product quality and reducing risks associated with seed availability. Based on fusion of morphogenic proto-plasts and genetic engineering to create hybrids, new approaches and priorities in tree improvement will emerge. Priorities will relate field tests for precise genetic adaptation to environmental constraints, to the nature and extent of the crop rotation, industrial needs, and public demand.

SPE may be applied to the Christmas tree market (e.g., blue, well-formed crowns) and to high-valued dwarf horticultural varieties in urban environments. Also useful would be the selection and cloning of trees suitable for the afforestation and reforestation of marginal lands and unproductive agricultural land. In this way, germplasm of value for fuel, fiber, structural materials, chemical feedstock, and possibly food through food chains (agroforestry) could be raised in zones where none existed before. However, it should be recognized that the introduction of precisely adapted trees could endanger rare plants on marginal soils.

In North American forestry, the softwoods will become the focus of many tree improvement programs [23]. This view is based on the projected 50-yr growth of hardwoods, which in the United States is tenfold greater than with softwoods. Improvement in softwoods is needed because conifers predominate in artificial reforestation programs in the United States and Canada.

New opportunities are available to study genetics and developmental biology for tree improvement. We can now search for the molecular bases of elite trees in ways never before possible. We have yet to explore gene expression unhindered by barriers of natural selection and evolution. Many new problems will emerge, some more serious than those currently encountered, largely because of our ignorance and increasing dependence on high technology. For this reason, we should also learn lessons of yet untouched forests before we engage and accept the new technologies emerging in forest biology.

REFERENCES

1. Abo El-Nil MM: Method for asexual reproduction of coniferous trees. U.S. Patent No. 4,353,184, 1982.
2. Allen GS, Owens JN: "The Life History of Douglas-Fir." Environment Canada, Forestry Service, Ottawa, 139 pp, 1972.
3. Bonga JM, Durzan DJ: "Cell and Tissue Culture in Forestry. 1. General Principles and Biotechnology. 2. Specific Principles and Methods: Growth and Development. 3. Case Histories: Gymnosperms, Angiosperms and Palms." Dordrecht: Martinus Nijhoff, 1987.
4. Camefort H: Fécondation et proembryogénèse chez les Abietacées (notion de neocytoplasme). Rev Cytol Biol Vég 32:253–271, 1969.
5. Canadian Forestry Service: Tree improvement in the boreal forest: Today and tomorrow. Proc 17th Mtg Can Tree Improv Assoc, Gander Newfoundland, Aug 27–30, 1979. Environment Canada, Ottawa, 1980.
6. Dandekar AM, Gupta PK, Durzan DJ, Knauf V: Genetic transformation and foreign gene expression in micropropagated Douglas-fir (*Pseudotsuga menziesii*). Bio/Technology 5:587–590, 1987.

7. Dogra PD: Variability in biology of flowering in blue pine provenances of northwestern Himalayas in relation to reproductive barriers and gene flow. In Proc Symp Flowering Physiology XVII IUFRO World Congress, Kyoto, Japan, 1981, Japan Forest Tree Breeding Assoc, Tokyo, 8–15.

8. Dogra PD: Seed sterility and disturbances in embryogeny in conifers with particular reference to seed testing and breeding in Pinaceae. Stud For Succica No. 45:1–97, 1967.

9. Drake JW, Glickman BW, Ripley LS: Updating the theory of mutation. Am Sci 71:621–630, 1983.

10. Durzan DJ: Ammonia: Its analogues, metabolic products and site of action in somatic embryogenesis. In Bonga JM, Durzan DJ (eds): "Cell and Tissue Culture in Forestry." Dordrecht: Martinus Nijhoff, Dr. W. Junk, Vol 2, pp 92–136, 1986.

11. Durzan DJ: Biotechnology and the cell culture of woody perennials. Forestry Chronicle 61:439–447, 1985.

12. Durzan DJ: Progress and promise in forest genetics. In "Paper Science and Technology, the Cutting Edge," Proc 50th Annu Conf, Appleton, WI, May 8–10, 1978: Institute of Paper Chemistry, pp 31–60, 1980.

13. Durzan DJ: Somatic embryogenesis and sphaeroblasts in conifer cell suspension. Proc V Int Congr Plant Cell and Tissue Culture, Tokyo, Japan, July 11–16, 1982, pp 113–114, 1982b.

14. Durzan DJ: Plant growth regulators in cell and tissue culture of woody perennials. Plant Growth Reguln 6:95–11, 1987.

15. Durzan DJ, Steward FC: Cell and tissue culture of white spruce and jack pine. Can Dept For Rural Dev Res Rep 24:30, 1968.

16. Durzan DJ, Steward FC: Morphogenesis in cell cultures of gymnosperms: some growth patterns. Invited paper, International Union of Forest Research Organizations. Section 22. Workshop held May 28–June 5, Helsinki, Finland, 20 pp. plus 8 plates. Abstr Comm Inst For Fenn 74(6):16 (1971), 1970.

17. Fraley RT, Rogers SG, Horsch RB: Genetic transformation in higher plants. CRC Crit Rev Plant Sci 4:1–46, 1986.

18. Gupta PK, Durzan DJ: Plantlet regenerations via somatic embryogenesis from subcultured callus of mature embryos of *Picea abies* (Norway Spruce). Rapid Comm Cell Biology, 22:685–688, 1986.

19. Gupta PK, Durzan DJ, Finkle BJ: Somatic polyembryogenesis in embryogenic cell masses of *Picea abies* (Norway spruce) and *Pinus taeda* (loblolly pine) after freezing in liquid nitrogen. Can J For Res 17:1130–1134, 1987.

20. Gupta PK, Durzan DJ: Somatic polyembryogenesis from callus of mature sugar pine embryos. Bio/Technology, 4:643–645, 1986.

21. Gupta PK, Durzan DJ: Biotechnology of somatic polyembryogenesis and plantlet regeneration in loblolly pine. Bio/Technology 5:147–151, 1987.

22. Gupta PK, Durzan DJ: Somatic embryos from protoplasts of loblolly pine proembryonal cells. Bio/Technology 5:710–712, 1987.

23. Haissig BE, Nelson ND, Kidd GH: Trends in the use of tissue culture in forest improvement. Bio/Technology 5:52–59, 1987.

24. Hakman I, Fowke LC, Von Arnold S, Eriksson T: The development of somatic embryos of *Picea abies* (Norway Spruce). Plant Sci Lett 38:53–59, 1985.

25. Kellison RC: Cone and seed harvesting from seed orchards. In Faulkner R (ed): "Seed Orchards." For Comm London Bull 54:101–107, 1975.

26. Krogstrup P: Embryo-like structures for cotyledons and ripe embryos of *Picea abies*. Can J For Res 16:664–668, 1986.

27. Litvay JD, Johnson MA, Verma DC, Einspahr DW, Weyrauch K: Conifer suspension culture medium development using analytical data from developing seeds. Inst Paper Chem Tech Pap Ser 115:1–17, 1981.
28. Longman KA, Dick JMP: Can seed orchards be miniaturized? In Proc Symp Flowering Physiology XVIII IUFRO World Congress, Kyoto, Japan, Forest Tree Breeding Assoc, Tokyo, 98–101, 1981.
29. Lower ARM: The North American assault on the Canadian forest. Ryerson Press, Toronto, 1938.
30. Nagmani R, Bonga JM: Embryogenesis in subcultural callus of *Larix decidua*. Can J For Res 15:1088–1091, 1985.
31. Norstog K: Experimental embryology of gymnosperms. In Johri BM (ed): "Experimental Embryology of Vascular Plants." Springer-Verlag, 25–51, 1982.
32. Schlarbaum SE: Cytogenetic manipulations in forest trees through tissue culture. In Bonga JM, Durzan DJ (eds): "Cell and Tissue Culture in Forestry." Dordrecht: Martinus Nijhoff, 1986, Vol 1, pp 330–352.
33. Silen RR: Nitrogen, corn, and forest genetics. USDA For Serv Gen Tech Rep PNW-137, 1972.
34. Singh H: "Embryology of gymnosperms." Enc Plant Anatomy, Vol 10, Part 2. Berlin: Gebrüder, Borntraeger, 1978.
35. Spiegel-Roy P, Vardi A: Citrus. In Ammirato PV, Evans DA, Sharp WR, Yamada Y (eds): "Handbook of Plant Cell Culture," Vol 3. 355–372, 1984.
36. Steward FC: "Growth and Organization in Plants." Reading, MA: Addison-Wesley, 1968.
37. Sweet GB, Krugman SL: Flowering and seed production problems—and a new concept of seed orchards. Proc 3rd World Consultation on Forest Tree Breeding CSIRO, Canberra, Vol 2, pp 749–759, 1978.
38. Timmis R: Factors influencing the use of clonal materials in commercial forestry. In "Crop Physiology of Forest Trees." 1986, pp 259–272.
39. Timmis R, Abo El-Nil M, Stonecypher RW: Potential genetic gains through tissue culture. In Bonga JM, Durzan DJ (eds): "Cell and Tissue Culture in Forestry," Vol 1. Dordrecht: Martinus Nijhoff, 1986, pp 198–215.
40. Ulrich JM, Michler RA, Finkle BJ, Karnosky DF: Survival and regeneration of American elm callus cultures after being frozen in liquid nitrogen. Can For Res 14:750–753, 1984.
41. Wilke-Douglas M, Perani L, Radke S, Bossert M: The application of recombinant DNA technology toward crop improvement. Physiol Plant 68:560–565, 1986.
42. Durzan DJ, Gupta PK: Somatic embryogenesis and polyembryogenesis in Douglas-fir cell suspension cultures. Plant Sci 52:229–235, 1987.
43. Durzan DJ: Metabolic phenotypes in somatic embryogenesis and polyembryogenesis. In Hanouver J. Keathly D (eds): "Genetic Manipulation of Woody Plants." Plenum Press, 1988, pp 293–312.
44. Durzan DJ: Somatic polyembroygenesis and plantlet regeneration in selected tree crops. Biotech and Genetic Eng Revs 6 (in press), 1988.

V. APPENDIX A: GLOSSARY

Auxins are natural and synthetic plant-growth regulators that promote cell elongation and the establishment of physiological states that promote embryogenesis.

Callus is a growth of unorganized, unconnected, or loosely connected plant cells normally produced from culturing an explant. An **embryogenic callus** represents calli that have the potential to produce embryos. This is distinct from a proliferating **embryonal-suspensor mass** (ESM), which is not considered a callus because daughter cells in the ESM repetitively yield somatic embryos as distinct from the random and nondescript populations of cells in callus which may not yield true-to-conifer type SPE.

Tissue culture is the process by which tissue excised from a donor plant is nourished and conditioned under aseptic conditions on a series of culture media to establish cultures "for maintenance" or to produce multiple plantlets genetically identical with the donor (micropropagation) and in some cases plantlets with aberrant phenotypes (somoclonal variants).

"Conifer-type" refers to one of the four types of proembryo development in conifers. This type of proembryogeny occurs in conifers and taxads and represents a basal plan for embryonic development. The terminology of conifer-type development is reviewed by Singh [34]. The terminology is used in describing stages in Figure 3.

Conversion represents the equivalent of the germination of a seed but refers to the establishment in soil of embryos that have been grown from somatic cells.

Cytokinins are natural and synthetic plant growth regulators that affect the organization of developing tissues and effect cells mainly by cell division.

An **embryogenic cell suspension** is a cell suspension derived from an embryonal-suspensor mass rather than callus.

Neocytoplasm forms the ground substance of a proembryo and is derived from the nucleoplasms of both the male and female nuclei at fertilization [4].

Parthenocarpy refers to the development of a seedless fruit or seed which lacks embryos. These processes are distinct and separate from the process of "cleavage polyembryogenesis" (see Figs. 3, 4).

Parthenogenesis is the production an embryo from a female gamete without the participation of a male gamete.

Polyembryony is the result of more than one embryo from a single egg cell, or from other sporophytic or gametophytic cells (in plants).

In **simple polyembryony** several egg cells develop from one megaspore and each is fertilized by a separate sperm or spore, or their development is parthenogenetic.

In **cleavage polyembryony** more than one embryo results by mitotic division (during each cleavage) of the zygote into two or more units, each developing into an embryo. The resultant embryos are "monozygotic" in origin and genetically identical and represent the new generation, as in *Pinus* sp.

In **sporophytic polyembryony** adventitious embryos arise by sporophytic budding from the nucellus and from the integument in flowering plants. The embryos are pseudogamic and usually identical with each other and with the mother plant, as in *Citrus*.

Proembryogeny encompasses the preliminary stages of development of the embryo. These stages occur before the elongation of the embryo associated with any given embryo. This is distinct from **early embryogeny** (all stages after elongation of the suspensor) and **late embryogeny** (establishment of the polar meristems (root and shoot) and the subsequent development of the embryo.

Sphaeroblast is a globular cellular mass of cells resembling a proembryo but derived from callus or cells with potential for meristematic growth. Sphaeroblast growth is distinguished by the internal production of precociously vascularized cells from a cambial-like meristem. Sphaeroblasts can polarize and resemble the early stage of embryonic development. Generally, however, they do not produce synchronized root and shoot development as in true-to-type zygotic or somatic polyembryogenesis. Sphaeroblast nodules are often confused with somatic embryos because of the similarity in outward appearance at early stages of development.

Suspensor is a group or chain of cells that is produced at one end of the developing proembryo and which serves usually to push the embryo into contact with food supply for nourishment.

Biotechnology in Agriculture, pages 83–96
© 1988 Alan R. Liss, Inc.

Virus Detection in Squash-Blots of Plants and Insects: Applications in Diagnostics, Epidemiology, and Breeding

Henryk Czosnek and Nir Navot

Department of Field and Vegetable Crops, Faculty of Agriculture, The Hebrew University of Jerusalem, Rehovot 76100, Israel

———◇◆◇———

———◇◆◇———

I. NEED FOR IMPROVED DIAGNOSTIC TOOLS

Viruses are among the major causes of crop losses. In order to predict and monitor virus epidemics, there is a need for the improvement or the development of rapid and sensitive detection procedures. Virus should be detectable in all plant tissues and in insect vectors (whenever relevant) as early as possible in the infection process. To be of practical and economical value, assays should be simple, reliable, cheap, and packageable in the form of a kit. These procedures should be readily usable not only by equipped laboratories, but also by the growers themselves.

The techniques for virus diagnostics which are currently available do not satisfy all these criteria. Widely used assays for virus contamination include the inoculation of extracts onto indicator plants and waiting (sometimes several weeks) for the development of typical disease symptoms [1]. More rapid is the use of immunological procedures [2] relying on the preparation of antibodies, usually against the virus coat protein, and their immunoabsorbtion by either anatomical sections (immunosorbent electron microscopy) [3] or by plant extracts (enzyme-linked immunosorbent assay—ELISA) [4,5]. The ELISA procedure is simple and sensitive, and much of the plant diagnostic work makes use of this technique, but there are many viruses which are difficult to purify in sufficient amounts to obtain high-quality antisera or which are poorly antigenic.

Nucleic acid hybridization is used more and more in diagnostics. Purified plant nucleic acids can be hybridized with probes prepared from viral nucleic acids [6]. Methods have been developed whereby the presence of viruses can be detected in clarified plant saps, avoiding the purification of nucleic acids and permitting the testing of a large number of samples in a relatively short time [7,8].

Although the trend for virus diagnostics is towards reducing to a minimum the number of steps in the assay, the techniques mentioned above are still time-consuming and labor-intensive; and tests have to be made in the laboratory. These are probably the reasons that use of commercially available diagnostic kits is not more widespread.

We present here a simple, specific, and rapid method for the detection of DNA and RNA viruses in plants and in insect vectors, based on the squashing of the plant or the insect material onto a nylon membrane, followed by hybridization with a virus-specific probe. Squash-blotting may be a breakthrough in plant-virus diagnostics, allowing the design of commercial kits.

II. DETECTION OF VIRUS IN SQUASH-BLOTS

The power of the technique is demonstrated here for a whitefly-transmitted geminivirus affecting tomato crops, the tomato yellow leaf curl virus (TYLCV). This virus causes a disease which affects cultivated tomato crops (*Lycopersicon esculentum* Mill.) in the Eastern Mediterranean basin and North Africa [9–11], leading to reduced yields. Symptoms similar to those of the TYLCV disease have been also described in Southeast Asia, Taiwan, Central Africa, and Mexico [12]. The virus is transmitted by the whitefly *Bemisia tabaci* Genn. [13] and is most effective in late summer and autumn. All tomato cultivars currently available are susceptible to the disease. We have recently isolated the TYLCV [14]. It is a single-stranded circular DNA geminivirus sharing many features with other viruses of the gemini group [recently reviewed in refs. 15,16]. The detection of RNA viruses with the squash-blot method was tested in tobacco mosaic virus (TMV)-infected tobacco plant tissues.

A. Detection of Virus Nucleic Acids in Leaf Squashes

Our goal was to eliminate most preparative steps on the tested material prior to its hybridization with specific probes. Our working hypothesis was that it should be feasible to detect viral nucleic acids on plant tissues squashed and permanently fixed onto a membrane (squash-blots), adapting procedures developed for bacterial colony hybridization [17], and for the visualization of highly repeated plant genomic DNA [18]. To make the viral nucleic acids available to the specific probes, we thought that we would have to precondition the squash-blots both to denude the viral DNA from its capsid and to protect the naked DNA from extruded plant nucleases.

Leaves from healthy and from TYLCV-infected tomato plants were directly squashed onto a nylon membrane (Hybond-N, Amersham, UK) using a hard object (pen, glass cylinder, etc.). Plant squashes were permanently fixed onto the membrane by either a 3-min irradiation on an ultraviolet table transilluminator or by a 3-min treatment in a microwave. Blots were then either incubated for 1 h at 50°C with 50 μg/ml proteinase K in 0.4% sodium dodecyl sulfate (SDS) prior to hybridization, or pretreated with alkali (0.5 N NaOH for 5 min at room temperature) before being neutralized; blots were also boiled for 10 min. Squash-blots were then hybridized with radiolabeled cloned TYLCV DNA. When we compared the hybridization results obtained after these treatments with those obtained with untreated blots, no remarkable differences were found. Therefore, it is possible to detect the presence of viral nucleic acids immediately after squashing plant material, without any previous treatment of the blots.

Very strong signals were obtained with TYLCV-infected plant tissues. Tissues from uninfected plants did not react at all (Fig. 1).

Sometimes, it is desirable to diagnose a disease very rapidly in order to design countermeasures (e.g., spraying insecticides, destroying infected plants). Usually, blot hybridization involves a minimum of 3 h of prehybridization and a minimum of 12 h of hybridization [19]. We have shortened the prehybridization period to 1 h without decreasing the sensitivity of the hybridization signal. Completely suppressing this step markedly increases nonspecific signals. Hybridization time could be reduced to 5 h without decreasing signals; but after 1 h of hybridization, the signal was reduced by 90%, which may sometimes still be adequate.

In summary, we are now using the following blotting and hybridization conditions: after squashing, the blot is illuminated for 2–5 min under a long-wave UV lamp, prehybridized for 1 h, and hybridized for 12 h. With a freshly radiolabeled probe, a 4-h exposure time is sufficient to obtain an adequate autoradiographic signal.

B. Specificity of Detection With the Viral Probe

The specificity of detection was tested by attempting to hybridize the TYLCV probe with squash-blots from tomato plants infected by other viruses. Leaves from tomato plants infected with either the cucumber mosaic virus (CMV) or the potato virus Y (PVY) failed to hybridize with the probe, as did leaves infected with the tobacco mosaic virus (TMV). Leaf squash-blots from an *Abutilon sellovianum* plant infected with the Abutilon mosaic virus (AbMV), another geminivirus [20], did not react with the TYLCV probe.

C. Distribution of Virus in the Plant Tissues

In some instances, especially in the case of systemic infections, it is of importance to monitor the spreading of the virus in the plant tissues. This can be easily and rapidly achieved with the squash-blot technique.

Like leaves, roots, flowers, and pollen grains can also be squashed. Stems slices can be stamped on the membrane. Fruits and seeds can be cut open prior to application onto the filter. The analysis of tissues from a TYLCV-infected plant indicates that viral nucleic acids are detectable in all the tissues and organs tested (Fig. 2). Moreover, this *in situ*-like hybridization can reveal the anatomical localization of the virus in these tissues. For example, hybridization of stem cross-sections indicates that viral sequences are

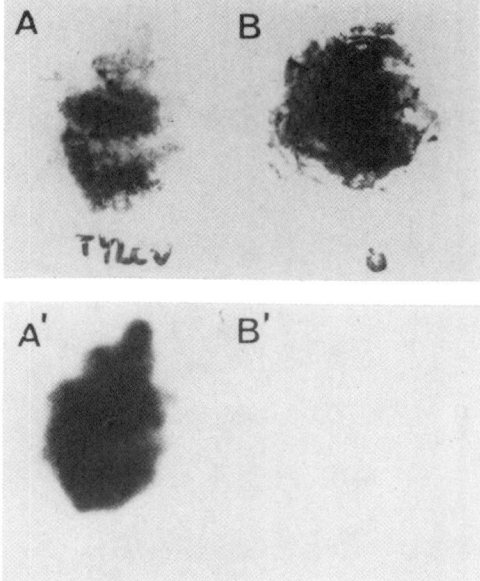

Fig. 1. Autoradiographic detection of TYLCV nucleic acids in a TYLCV-infected tomato leaf squash. Photograph of leaf squash-blots (**upper row**) from TYLCV-infected (**A**) and healthy (**B**) tomato plants, and the corresponding autoradiograms (**lower row, A′, B′**), following hybridization with the cloned TYLCV-DNA probe. Tomato plants (*L. esculentum*) were inoculated in the greenhouse with viruliferous whiteflies, after an acquisition period of 18 h [13]. The blots were prehybridized for 3 h and hybridized for 16 h [19] with the probe radiolabeled with ^{32}P by nick-translation [22]. The blots were washed twice for 20 min in $1 \times$ SSC (150 mM NaCl-15 mM trisodium citrate) and exposed for 16 h to a Kodak X-Omat film. The virus-specific probe was made from the presumptive viral replicative form (RF) of the TYLCV genome isolated from a TYLCV-infected tomato plant and cloned in a plasmid as follows: DNA isolated from an infected tomato plant [23] was electrophoresed in a 1.5% NA-agarose (Pharmacia Biotechnology, Uppsala, Sweden) gel [17] and a longitudinal segment was transferred [24] to a Hybond-N membrane (Amersham, UK) and hybridized [19] to nick-translated [22] TYLCV genomic DNA [14]. TYLCV-related DNA species were localized and the presumptive double-stranded replicative form (RF) of the TYLCV genome was isolated by electroelution from the agarose gel. Full-length TYLCV double-stranded DNA was cloned in the AccI site of the pTZ18 cloning vector polylinker (Pharmacia Biotechnology, Uppsala, Sweden), following linearization with Hpa II. A 1.7-kb HpaII/SphI TYLCV DNA fragment (encoding most of the coat protein) was subcloned in the same vector (designated pTZ18-hs11) and used as a probe. For squash-blot detection, we used either tomato plants (*L. esculentum*) inoculated in the greenhouse with viruliferous whiteflies or cultivars (M82) naturally infected in the field.

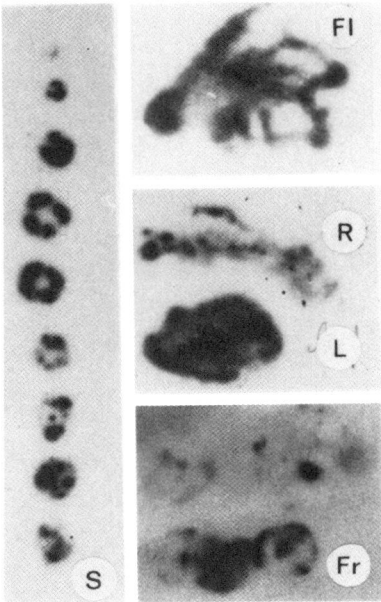

Fig. 2. Autoradiographic detection of TYLCV DNA sequences in squash-blots of TYLCV-infected tomato plant tissues. A tomato plant grown in the field (*L. esculentum*, cv. M82), with characteristic disease symptoms, was sampled; different tissues were squash-blotted and hybridized with the TYLCV-specific probe. **S**, stem slices from top (**up**) to crown (**down**); **Fl**, flower; **R**, roots; **L**, leaf; **Fr**, fruit cut open (note the strong signal given by seeds).

preferentially found in the vascular system (Fig. 2). In leaves, the virus is mainly detected in the veins (Fig. 3).

Since it is known that the TYLCV disease is not transmitted by seeds, it was of interest to test seeds prepared from infected tomato plants. Considerable amounts of virus sequences were found associated with the seeds in cut-open fruits (Fig. 2) as well as in freshly prepared seeds. The storage of seeds for 1 mo or more made the virus barely detectable. Therefore, the fact that the disease is not transmitted by seeds may be due to the use of long-term-stored seeds.

D. Squash-Blots Can Be Used to Detect DNA as Well as RNA Viruses

Squash-blots can be used, with minor modifications, to detect RNA viruses. Leaves from healthy and TMV-infected tobacco plants were

LEAF SQUASH-BLOT

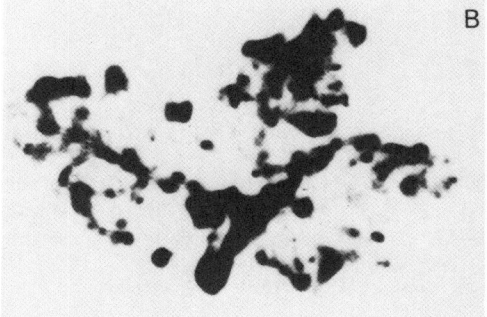

Fig. 3. Autoradiographic visualization of virus DNA sequences in TYLCV-infected tomato leaf. **A:** Photograph of leaf squash-blot. **B:** Corresponding autoradiogram after hybridization with TYLCV-specific probe.

squashed onto a membrane previously saturated with SDS and proteinase K (to protect the viral genome from RNAases) and dried. A strong and specific signal was obtained following hybridization with a reverse transcript of TMV-RNA (Fig. 4).

E. Detection of Virus in Squashes of Insect Carriers

Since many plant diseases are spread by insects, it is crucial to be able to monitor viruliferous populations. We show here that virus concentrations in insects can be determined at the level of the individual carrier.

Whiteflies kept for 18 h on TYLCV-infected or healthy Datura plants

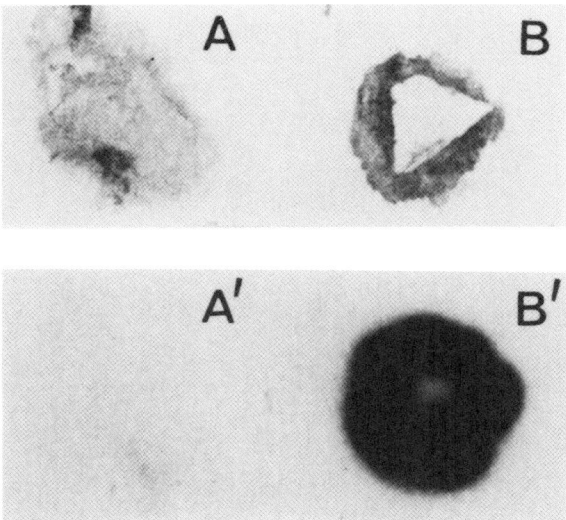

Fig. 4. Detection of tobacco mosaic virus (TMV) RNA sequences in squash-blots of TMV-infected tobacco leaves.

Upper panel: Photograph of squash-blot from healthy (**A**) and infected (**B**) leaves. **Lower panel:** Corresponding autoradiogram (**A'**, **B'**, respectively) after hybridization with TMV-specific probe. Tobacco leaves (*Nicotiana tabacum*, cv. Samsun) were squash-blotted on a membrane presoaked in 0.5% sodium dodecyl sulfate (SDS) and 100 μg/ml proteinase K. The blot was hybridized for 12 h with a TMV-RNA reverse transcript [25], washed, and autoradiographed as described in Figure 1.

(*Datura stramonium*) (a commonly used test plant for TYLCV) were individually squashed onto the nylon membrane and hybridized with the viral probe. The autoradiographic analysis indicated that viral sequences are detectable in each one of the viruliferous whiteflies (Fig. 5). However, the intensity of the hybridization signals showed considerable variability from one individual to the other, ranging from an estimated 1 million to 100 million copies of the viral genome. No hybridization was obtained with nonviruliferous whiteflies.

III. ADVANTAGES OF SQUASH-BLOTS OVER OTHER VIRUS DETECTION TECHNIQUES

The squash-blot technique for virus detection provides the tool needed for rapid, large-scale diagnostics. Some of the advantages of the method are discussed below.

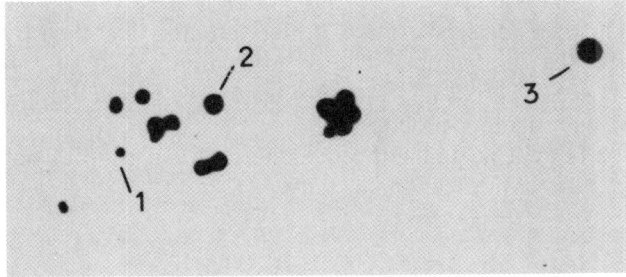

Fig. 5. Detection and quantification of TYLCV-DNA sequences in viruliferous whiteflies. Female whiteflies were kept on TYLCV-infected Datura plants for an acquisition period of 18 h in an insect-proof greenhouse [13], collected, and immediately frozen at −20°C. Viruliferous and control whiteflies were individually squashed onto a membrane and hybridized with the viral probe as described in Figure 1. Quantification was done by comparing the hybridization signals with those obtained with known amounts of viral DNA (in a range equivalent to 50–0.5 million genome copies). Virus sequences were detectable in each one of the viruliferous whiteflies, but concentrations varied. For example, the whitefly marked **1** contained about one million copies of the viral genome; whitefly **2**, about 15 million copies; and whitefly **3**, more than 50 million copies.

A. Simplicity

There is no need to prepare DNA or saps from test material, only to squash tissues on a membrane. Hundreds of samples can be processed daily by untrained personnel. Squash-blots from as little as 50 mg of any plant tissue (a few mm^2) are sufficient for detection.

B. Sensitivity and Specificity

The high sensitivity of the method is demonstrated by the ability to detect virus nucleic acids in a single whitefly. Specificity is dictated by the viral probe.

C. Rapidity

Results can be obtained as soon as 18 h after sampling, depending on the concentration of virus in infected tissues.

D. Stability

Squash-blots are surprisingly stable. Leaf squash-blots were stored under several extreme conditions: blots were 1) immediately processed after squashing; 2) kept at room temperature for 5 days; 3) frozen and thawed several times; 4) boiled for 2 min; and 5) stored at 37°C for 5 days at 100% humidity. Only the materials stored at high humidity lost their hybridization capacity with the viral probe. The hybridization signal sustains stringent washes (0.1 × SCC at 70°C). Squash-blots can be boiled to remove a probe and utilized with another probe.

Squash-blots kept at ambient temperatures for 6 mo did not show any reduction in their hybridization capacity.

IV. POTENTIAL USES OF SQUASH-BLOTS

A. Epidemiological Studies

Many viral diseases, like the whitefly-borne TYLCV, are spread by insects. The TYLCV disease is acute during the tomato-growing season, namely, in summer and autumn. It is believed that between growing seasons, plants near tomato fields (e.g., *Cynanchum acutum*) are the natural hosts of the virus and serve as a source for its spreading [21]. Identifying these plants by the squash-blot technique may be the basis of a sanitation program aimed at eradicating potential virus hosts.

It has been found that there is a correlation between the size of the whitefly population and the disease incidence [21], but it is not known whether there is a critical virus concentration in insects below which the whitefly is harmless. The squash-blot technique makes it possible to identify plant hosts and also to determine the concentration of virus in the insect carriers.

It is now feasible to predict TYLCV epidemics by sampling whiteflies from the beginning of the growing season, possibly 2 or 3 times a week, in or near tomato fields, and to test them for the presence of the virus. A rise in virus concentration may be the signal for starting or intensifying the spraying of insecticides. Since whitefly populations migrate in an unpredictable path, monitoring may have to be done on a regional, national, or international scale.

B. Mapping Viral Diseases Worldwide

Squash-blotted samples are stable and can be kept for at least 6 mo at ambient temperatures before analysis. Therefore, an unlimited number of plants with suspect disease symptoms (a few mm^2 per sample) can be collected and mailed from one country to another. Blots can either be analyzed on the spot or mailed to a central diagnostic facility, which may be private, governmental, or international. Climatic conditions such as humidity, dryness, or large differences in temperature do not significantly degrade the material to be tested from the moment a squash is made till it can be analyzed.

In collaboration with Drs. Laterrot and Bordat (INRA, France), we are mapping the occurrence of the TYLCV disease worldwide. Squash-blots from plants presenting typical TYLCV symptoms were sent to us for analysis from Senegal; the results indicated the presence of TYLCV in five different regions of this country.

C. Breeding for Virus Resistance

We are using squash-blots for screening a large number of plants generated during a breeding program for TYLCV resistance. All the cultivated tomato species (*L. esculentum*) are susceptible to the yellow leaf curl disease. Efforts are being made to introgress disease resistance traits found in some wild tomato species into tomato cultivars. One of the main reasons for the slow progress in these breeding programs was the absence of a unequivocal screening test for resistance to the TYLCV.

In three growing seasons, accessions of two wild tomato species, *L. hirsutum* (LA 1777) and *L. pinnellii* (LA 716), have not developed the typical viral-disease symptoms when grown in fields where *L. esculentum* (cv. M82) were totally infected. No virus was detected in leaf squashes from these wild tomato species, whereas virus was detected in squashes from the cultivated tomato. In the F1 generation resulting from crosses of the wild species to the cultivated tomato (LA 716 × M82), both disease symptoms and viral sequences were detected. In the F1 LA 1777 × M82, virus sequences could be detected, but the plant was symptomless (Fig. 6). Therefore, the latter plant was tolerant to viral infection; the virus was able to replicate but did not express the disease phenotype. Using the squash-blot technique, we are now screening susceptible and resistant plants at the seedling stage, allowing selection at an early stage with the advantages of accuracy, space, and economy of time and manpower.

D. Quality Control

Growers buying seeds or seedling from commercial firms are interested in getting virus-free products. Common practice is that firms keep a significant fraction of their output for future biological tests, in case clients claim that the product was contaminated. Squash-blots can serve as an unequivocal record for the products sold, since either a single seed or a small piece of leaf is necessary for diagnostics.

V. CONCLUSIONS

Squash-blots for the detection of viruses are a versatile instrument with many potential applications in various aspects of agriculture, as discussed

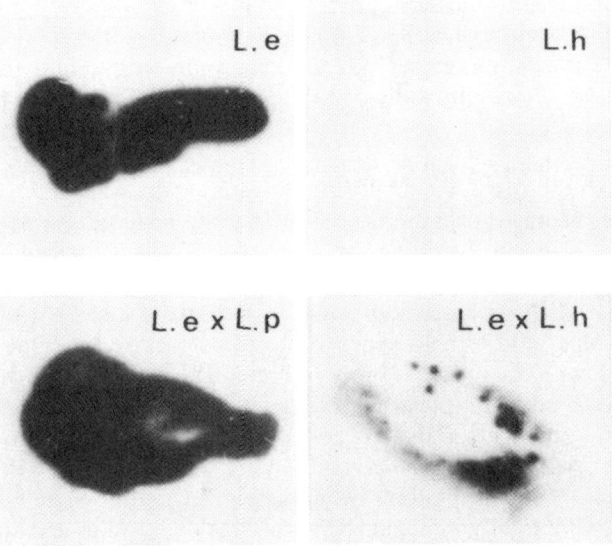

Fig. 6. Autoradiographic detection of TYLCV DNA sequences in hybrids between cultivated and wild tomato species. Hybrids F1 were made between the cultivated tomato *L. esculentum* cv. M82 and accessions of the two wild species, *L. hirsutum* (LA 1777) and *L. pennellii* (LA 716), which were symptomless when grown in fields where *L. esculentum* cultivars were totally infected. Blots containing leaf squashes from parents and F1 plants grown together in a TYLCV-infected tomato field were hybridized with the TYLCV-specific probe. **L.e.**, diseased *L. esculentum*; **L.h.**, symptomless *L. hirsutum*; **L.e.** × **L.p.**, hybrid *L. esculentum* × *L. pennellii* with disease symptoms; **L.e.** × **L.h.**, symptomless hybrid *L. esculentum* × *L. hirsutum*.

above. Viral nucleic acids can be specifically and easily detected in many tissues and organs of the infected plant: leaf, root, stem, flower, seed, and fruit. Virus can also be detected in insect carriers, at the level of the single individual.

The method is sensitive and simple: samples do not have to be prepared for analysis in a laboratory: a simple squash of a small piece of leaf on a membrane is sufficient. Many samples can be processed in a short time by unexperienced personnel. The squash-blots are very stable; they can be either processed immediately, stored for months, or mailed to a diagnostic facility. These features make it possible to develop commercially available diagnostic kits. For this, we are now developing a system for virus detection in squash-blots using chromogenic nonradioactive probes (e.g., Chemiprobe,

Orgenics, Yavne, Israel). This should make the user completely independent of laboratory facilities.

ACKNOWLEDGMENTS

Whiteflies and greenhouse-infected tomato plants were provided by Dr. Antignus (Agriculture Research Organization, Bet-Dagan, Israel). TMV-infected and healthy tobacco plants, and TMV-RNA were from Dr. Sela (The Faculty of Agriculture, Rehovot). We thank Sarah Ovadia for excellent technical assistance. This work was supported by BARD grant I-1110-86.

REFERENCES

1. Hamilton RI, Edwardson JR, Francki RIB, Hsu HT, Hull R, Koenig R, Milne RG: Guidelines for the identification and characterization of plant viruses. J Gen Virol 54:223, 1981.
2. Clark MF, Bar-Joseph M: Enzyme immunoabsorbent assays in plant virology. In Maramorosch K and Koprowski H (eds): "Methods in Virology." New York: Academic Press, 1984, Vol 7, p 51.
3. Roberts IM, Harrison BD: Detection of potato leafroll and potato mop-top viruses by immunoabsorbent electron microscopy. Ann Appl Biol 93:289, 1979.
4. Engvall E, Perlman P: Enzyme-linked immunoabsorbent assay (ELISA)—quantitative assay of immunoglobulin G. Immunochemistry 8:871, 1971.
5. Van Regenmortel (ed): "Serology and Immunochemistry of Plant Viruses." New York: Academic Press, 1982.
6. Symons RH: Diagnostic approaches for the rapid and specific detection of plant viruses and viroids. In Kosuge T, Nester WE (eds): "Plant-Microbe Interactions. Molecular and Genetic Perspectives." London: MacMillan, 1:93, 1985.
7. Owens RA, Diener TO: Sensitive and rapid diagnosis of potato spindle tuber viroid disease by nucleic acid hybridization. Science 213:670, 1981.
8. Baulcombe DC, Flavell RB, Boulton RE, Jellis GJ: The use of cloned hybridisation probes to detect viral infection in a potato breeding programme. In Lea PJ, Stewart GR (eds): "Annual Proceedings of the Phytochemical Society of Europe," Oxford: Clarendon Press, 23:183, 1984.
9. Cohen S, Harpaz I: Periodic, rather than continual acquisition of a new tomato virus by its vector, the tobacco whitefly (Bemisia tabaci Gennadus). Ent Exp Appl 7:155, 1964.
10. Al-Musa A: Incidence, economic importance, and control of tomato yellow leaf curl in Jordan. Plant Dis 66:561, 1982.
11. Cherif C, Russo M: Cytological evidence of the association of a geminivirus with the tomato yellow leaf curl disease in Tunisia. Phytopathol Z 108:221, 1983.
12. Makkouk KM, Laterrot H: Epidemiology and control of tomato yellow leaf curl virus. In Plumb RT, Thresh JM (eds): "Plant Virus Epidemiology." Oxford: Blackwell, 1983, p 315.
13. Cohen S, Nitzany FE: Transmission and host range of the tomato yellow leaf curl virus. Phytopathology 56:1127, 1966.

14. Czosnek H, Ber R, Antignus Y, Cohen S, Navot N, Zamir D: Isolation of the tomato yellow leaf curl virus—a geminivirus. Phytopathology (in press).
15. Howarth A J: Geminiviruses, the plant viruses with single-stranded DNA genomes. In Setlow JK, Hollaender A (eds): "Genetic Engineering." New York: Plenum, 8:85, 1986.
16. Lazarowitz SG: The molecular characterization of geminiviruses. Plant Mol Biol Rep 4:177, 1987.
17. Maniatis T, Fritsch EF, Sambrook J: "Molecular cloning. A Laboratory Manual." Cold Spring Harbor, NY: Cold Spring Harbor Laboratories, 1982.
18. Hutchinson J, Abbott A, O'Dell M, Flavell RB: A rapid screening technique for the detection of repeated DNA sequences in plant tissues. Theor Appl Genet 69:329, 1985.
19. Carmon Y, Czosnek H, Nudel U, Shani M, Yaffe D: DNAase I sensitivity of genes expressed during myogenesis. Nucleic Acids Res 10:3085, 1982.
20. Abouzid A, Jeske H: The purification and characterization of gemini particles from Abutilon mosaic virus infected Malvaceae. J Phytopathol 115:344, 1986.
21. Cohen S, Keren J, Harpaz I, Bar-Joseph R: Studies of the epidemiology of a whitefly-borne virus, tomato yellow leaf curl virus, in the Jordan valley. Phytoparasitica 14:158, 1986.
22. Rigby PWJ, Dieckman M, Rhodes C, Berg P: Labelling deoxyribonucleic acid to high specific activity *in vitro* by nick translation with DNA polymerase I. J Mol Biol 113:237, 1977.
23. Taylor B, Powell A: Isolation of plant RNA and DNA. BRL Focus 4:4, 1982.
24. Southern EM: Detection of specific sequences among DNA fragments separated by gel electrophoresis. J Mol Biol 98:503, 1975.
25. Hull R: Purification, biophysical and biochemical characterisation of viruses with especial reference to plant viruses. In Mahy BWJ (ed): "Virology, a Practical Approach." Oxford: IRL Press, 1985, p 1.

Biotechnology in Agriculture, pages 97–140
© 1988 Alan R. Liss, Inc.

Alkaloid Production by Plant Cell Cultures

C.A. Hay, L.A. Anderson, M.F. Roberts, and J.D. Phillipson

Department of Pharmacognosy, The School of Pharmacy, University of London, London WC1N 1AX, England

Plants continue to be the major source of 25% of all prescription medicines in the United States and provide many of the raw materials used extensively by the flavor and fragrance industries [1]. Alkaloids are one of the most important groups of pharmacologically active principles found in plants, and about 30 of them are used medicinally for a wide range of pharmacological effects (Table I). While synthetic routes to many of the alkaloids are well established, economically viable processes exist for only a few of the medicinal alkaloids in common use, e.g., caffeine, ephedrine [2].

Over the last decade, the potential of plant cell cultures as an alternative means of producing commercially important secondary metabolites has been the focus of intense investigation [3–7]. Despite the initial optimism, few commercial successes have been achieved, the chief constraints being that

TABLE I. Pharmacologically Important Alkaloids

Pharmacological effect	Alkaloid
Analgesic	Morphine, codeine
Antiamoebic	Emetine
Antiarrhythmic	Quinidine, ajmaline
Antibacterial	Berberine
Anticancer	Vinblastine, vincristine, harringtonine, camptothecin
Anti-inflammatory	Colchicine
Antimalarial	Quinine
Antitussive	Glaucine, noscapine
Act on autonomic nervous system	Physostigmine, pilocarpine, atropine, hyoscyamine, ephedrine, nicotine, scopolamine
Act on central nervous system	Reserpine, caffeine
Hypotensive	Reserpine, rescinnamine, protoveratrines A and B
Local anesthetic	Cocaine
Muscle relaxant	Tubocurarine, papaverine, theophylline
Vasodilator	Vincamine

cultured cells tend to produce inferior yields of the desired secondary metabolites when compared to the whole plant. Nevertheless, it should be pointed out that more than 30 compounds are known to accumulate in cultures at levels higher than that of the plant, and these include several alkaloid-producing species such as *Catharanthus roseus* (ajmalicine and serpentine) and *Coptis japonica* (berberine) [4,8,9].

Although few high-yielding cultures have been developed, more than 200 alkaloids, including several novel compounds, have been successfully identified in plant cell cultures, albeit mostly in small amounts (Tables II–VIII). Hence, it is apparent from the relatively small number of species which have been investigated to date that there is a considerable biosynthetic capability for the production of alkaloids. The use of plant cell cultures, in recent years, has contributed greatly to our understanding of the biosynthetic path-ways of a number of alkaloids, particularly those of the tryptophan-derived indole group and from the tyrosine-derived isoquinoline group [10]. Undoubtedly, the use of tissue culture techniques will provide further insight into those factors which regulate alkaloid production in vitro.

The aim of this chapter is to review our knowledge of alkaloid production by plant cell cultures and to consider three main aspects—namely, the range of chemical types produced, their biosynthesis, and their localization/compartmentalization within cultures. Previous reviews dealing with the production of alkaloids by plant cell cultures have been published [11–13], and this chapter concentrates mainly on literature published from 1983 onwards.

I. ALKALOIDS PRODUCED BY PLANT CELL CULTURES

It is evident from the scientific literature that a wide range of alkaloids has been detected in cell cultures involving almost all of the main alkaloid groups, and these findings are summarized in Tables II–VIII [14–173]. A great deal of interest has focused on those alkaloids containing indole and isoquinoline moieties, and these represent the two major groups of plant alkaloids.

A. Tryptophan-Derived Alkaloids

More than 100 tryptophan-derived indole alkaloids have been identified from plant cell cultures, some being noniridoid types, e.g., β-carbolines, canthinones; but more than 90% of them belong to the iridoid-derived group (Table II). Representatives of almost all of the main structural types of iridoid-derived indoles, such as corynanthean, aspidospermatan, ibogan, strychnan, plumeran, eburnan, vallesiachotaman, apparicine, and indoloisoquinolines, have been found. Surprisingly, this large number of alkaloids has resulted from studies involving only 14 species representing 10 genera, viz., *Alstonia, Camptotheca, Catharanthus, Cinchona, Ochrosia, Picralima, Rauwolfia, Rhazya, Stemmadenia, Tabernaemontana,* and *Voacanga,* which apart from *Cinchona* (Rubiaceae) and *Camptotheca* (Nyssaceae), belong to the Apocynaceae.

Interest in species of the Apocynaceae stems from the pharmacological activities of many of the indole alkaloids which occur in the family. Not surprisingly, *Catharanthus roseus* has been the main focus of research into the use of plant cell cultures as an alternative source of the expensive, clinically useful alkaloids vincaleukoblastine and vincristine (Fig. 1). Despite considerable scientific endeavor, it would appear from the literature available that the production of these particular dimeric indole alkaloids has proved to be an elusive goal. However, two short communications published in 1986 indicate that both of these alkaloids are obtainable in low yields from *C. roseus* callus cultures [174,175]. Cell cultures of *C. roseus* have proved to be an excellent source of monomeric indole alkaloids (Table II); 45 alkaloids have been identified, ajmalicine and serpentine being the major products. Despite the lack of success in the production of vincaleukoblastine and vincristine, the related dimeric alkaloid, isoleurosine, has been tentatively identified in *C. roseus* cell suspension cultures [52]. The capability of plant cell suspension cultures to produce indole alkaloids has been demonstrated by the isolation of voafrine A and B (Fig. 1) from the related genus *Voacanga* [49] and more recently of 3-R/S-hydroxyconodurine and monogagaine from *Tabernaemontana elegans* [45]. The latter alkaloid has not been

TABLE II. Tryptophan-Derived Alkaloids From Plant Cell Cultures

Structural type	Alkaloid	Source [reference]
Indoles		
Noniridoid		
β-Carbolines	Harman, harmine, harmalol, harmol, harmaline, ruine	*Peganum harmala* [14,15]
Canthinones	Canthin-6-one, 1-hydroxy-canthin-6-one, 1-methoxy-canthin-6-one	*Ailanthus altissima* [16,17]
Iridoid[a]		
Corynantheans	Ajmalicine, akuammigine, akuammiline, alstonine, desacetylakuammiline, dihydrositsirikin, 3-epiajmalicine, 21-hyroxycyclolochnerin, 10-hydroxydes-acetylakuammiline, 7-hydroxyindolenine-ajmalicine, 3-isoajmalicine, isositsirikin, 3-iso-19-epicathenamine, 16-R-19, 20-E-isositsirikin, 16-R-19,20-Z-isositsirikin, pleiocarpamine, pseudoindoxyajmalicine, serpentine, sitsirikin, strictosidine lactam, tetrahydroalstonine, yohimbine	*Catharanthus roseus* [20–34]
	Alstonine, serpentine, akuammigine, sitsirikin, isositsirikin, 3-Iso-19-epicathenamine	*Catharanthus ovalis* [24]
	Quinamine	*Cinchona ledgeriana* [35]
	Quinamine, cinchonamine, 10-methoxy-cinchonamine	*C. pubescens* [36]
	Cathenamine, isoreserpiline, pleiocarpa-mine, reserpiline, tetrahydroalstonine	*Ochrosia elliptica* [37–39]
	Ajmalicine, ajmaline, alstonine, 3-epiajmaline, 3-isoajmaline, 3-iso-ajmalicine, 17-O-acetylajmaline, 17-O-acetylnorajmaline, raucaffricine, reserpine, sarpagine, serpentine, vinorine, vomilenine, yohimbine	*Rauwolfia serpentina* [41–43]
	Reserpine	*Alstonia constricta* [50,51]

TABLE II. Tryptophan-Derived Alkaloids From Plant Cell Cultures (continued)

Structural type	Alkaloid	Source [reference]
	Akuammidine, strictosidine lactam	*Rhazya stricta* [44]
	Geissoschizol, isositsirikin, tabernae-montanine, vobasine, vobasinol	*Tabernaemontana elegans* [45]
Aspidospermatans	19-acetoxy-11-hydroxytabersonine, 19-acetoxy-11-methoxytabersonine horhammercine, horhammerinine, 19-hydroxy-11-methoxytabersonine, 20-hydroxytabersonine, lochnericine, lochnerinine, minovincine, taber-sonine, vindoline, desacetylvindoline	*Catharanthus roseus* [21–31]
	Tabersonine, lochnericine, horhammercine	*C. ovalis* [24]
	1,2-dehydroaspidospermidine, tabersonine, vincadifformine, tetrahydrosecodine	*Rhazya stricta* [44]
	Tabersonine	*Stemmadenia tormentosa* [46]
	Lochnericine, minovincinine, tabersonine	*Voacanga africana* [46]
Ibogans	Catharanthine	*Catharanthus roseus* [23,24,26,27,29,30] *C. ovalis* [24]
	Coronaridine	*Stemmadenia tormentosa* [46]
	Catharanthine, coronaridine	*Tabernaemontana divaricata* [47]
	Coronaridine, 3-R/S-hyroxycoronaridine, 3-R/S-hydroxyisovoacangine isovoacangine, 3-oxoisovoacangine	*T. elegans* [45]
Strychnans	Akuammincine, vinervine	*Catharanthus roseus* [22–24]
	Akuammicine	*C. ovalis* [24]
	Norfluorocurarine	*Ochrosia elliptica* [39]
	Akuammicine, norfluorocurarine Condylocarpine, norfluorocurarine, tubotaiwine, vinervine	*Rhazya stricta* [44] *Stemmadenia tormentosa* [46]
	Tubotaiwine, vinervine	*Tabernaemontana divaricata* [47]
	Tubotaiwine	*T. elegans* [45] *T. iboga* [47]

TABLE II. Tryptophan-Derived Alkaloids From Plant Cell Cultures (continued)

Structural type	Alkaloid	Source [reference]
Plumerans	Vindolinine, 19-epivindolinine, 20-epivindolinine, 20-epivindolinine-N_b oxide, vindolinine-N_b oxide	*Catharanthus roseus* [21,22,24–30]
	Vindolinine, epivindolinine	*C. ovalis* [24]
Eburnans	Eburnamine, eburnamonine	*Rhazya stricta* [44]
Vallesiachotamans	Vallesiachotamine, isovallesiachotamine	*Catharanthus roseus* [21,22,26,28,29,31]
	Vallesiachotamine, isovallesiachotamine	*Rhazya stricta* [44]
Appancines	Apparicine	*Catharanthus ovalis* [24]
	Apparicine, epchrosine, conoflorine	*Ochrosia elliptica* [39,40]
	Apparicine, pericine	*Picralima nitida* [48]
	Apparicine, conoflorine	*Tabernaemontana divaricata* [47] *T. elegans* [45]
	Conoflorine	*T. iboga* [47]
	16-hydroxy-16,22-dihydroapparicine	*T. elegans* [45]
Indoloisoquinolines	Ellipticine, 9-methoxyellipticine	*Ochrosia elliptica* [37,38]
Dimeric indoles	Isoleurosine	*Catharanthus roseus* [52]
	3-R/S-hydroxyconodurine, monogagaine	*Tabernaemontana elegans* [45]
	Voafrine A and B	*Voacanga africana* [49]
Quinolines	Quinine, quinidine, cinchonine, cinchonidine, dihydroquinine, dihydroquinidine, dihydrocinchonine, dihydrocinchonidine	*Cinchona ledgeriana* [35,53–62,63[b]] *C. pubescens* [36,53–56, 59,64,65]
	Cinchoninone	*C. ledgeriana* [35]
	Camptothecin	*Camptotheca accuminata* [66–68]

[a]Skeletal types of iridoid alkaloids are classified according to [18,19].
[b]Culture transformed with *Agrobacterium tumefaciens*.

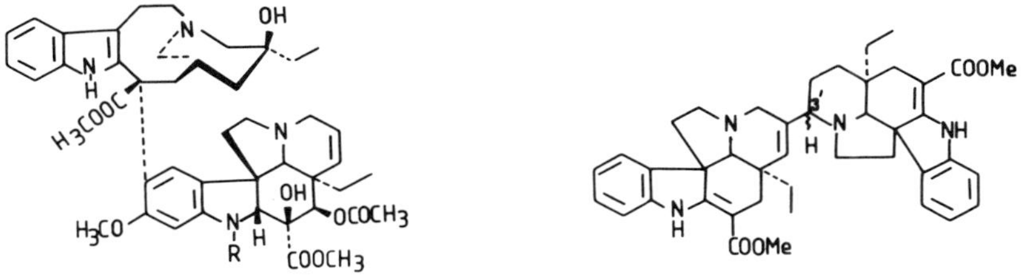

vinblastine   R = CH₃
vincristine   R = CHO

voafrine A   C-3'H α configuration
voafrine B   C-3'H β configuration

Fig. 1.   Vinblastine, vincristine, and voafrine A and B.

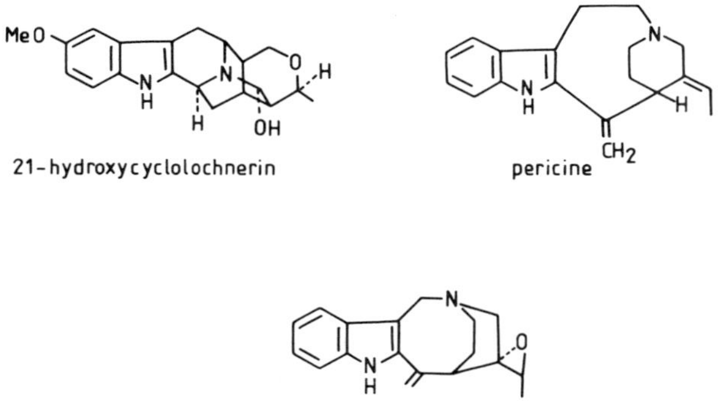

21-hydroxycyclolochnerin

pericine

Epchrosine

Fig. 2.   Epchrosine, 21-hydroxycyclolochnerin, and pericine.

identified in the whole plant but it has been found in other species of *Tabernaemontana* [45].

Studies with indole alkaloids have demonstrated the excellent potential of plant cell cultures to produce compounds not yet detected in the corresponding whole plant, but in some instances, compounds not previously described as natural products have also been isolated. Novel monomeric indole alkaloids have been reported from cell cultures of a number of species including 21-hydroxycyclolochnerin from *Catharanthus roseus* [34] and epchrosine from *Ochrosia elliptica* [39,40] (Fig. 2). The use of opiate binding studies, in conjunction with the isolation of active compounds from *Picralima nitida* cell cultures, has led to the identification of two alkaloids with opioid activity, one of which, pericine, is a novel alkaloid [48] (Fig. 2).

Apparicine (pericalline) was identified as the other CNS-active indole alkaloid, and it has not previously been reported from *Picralima nitida*, although it is a constituent of other Apocynaceous plants.

The alkaloids present in *Rauwolfia serpentina* cell cultures have been extensively investigated, and some 12 indole alkaloids have been identified (Table II), including ajmalicine, ajmaline, alstonine, reserpine, sarpagine, serpentine, vomilenine, vinorine, and yohimbine [41–43]. The glycoalkaloid raucaffricine, originally characterized as vomilenine galactoside, has had its structure revised to vomilenine β-D-glucoside [176]. Raucaffricine has been shown to be the major alkaloid of *R. serpentina* cell suspension cultures when alkaloid production medium was used [42]. Previously, raucaffricine has been reported as a constituent of *R. caffra* plants [177–179], but this is the first report of its occurrence in *R. serpentina*.

Cell cultures of *Rhazya stricta* have also proved to be rich sources of indole alkaloids, and 11 alkaloids have been identified, representing 5 of the main structural types of indole alkaloids—namely, corynanthean, aspidospermatan, strychnan, eburnan, and vallesiachotaman (Table II) [44]. These alkaloids, with the exception of akuammicine, are typical of the genus *Rhazya*. Examples of cultures producing alkaloids which are not typical of the parent plant include 3-oxoisovoacangine from *Tabernaemontana elegans* cultures [45] and reserpiline from *Ochrosia elliptica* cultures [38]. Although reserpiline has not been found in differentiated plants of *O. elliptica*, it has been detected as a minor alkaloid of *O. balansae* [38].

Indole alkaloids obtained from callus of *Cinchona* species (Rubiaceae) include quinamine from *C. ledgeriana* [35] and *C. pubescens* [36]; cinchonamine and 10-methoxycinchonamine have been isolated from callus of the latter species [36]. Some eight associated tryptophan-derived quinoline alkaloids, including quinine and quinidine, have been obtained from *Cinchona* callus and suspension cultures (Table II) [35,36,53–65]. The levels of quinine and quinidine have been low in comparison to the bark of the whole plant, and in some instances the nonmethoxylated analogs, cinchonidine and cinchonine, have proved to be the major alkaloids of cultures [35,36,53–65]. In a recent study, *C. ledgeriana* cultures transformed with *Agrobacterium tumefaciens* have been described [63]. The transformed cultures are capable of growing and producing quinoline alkaloids without the addition of phytohormones. However, the type and relative amounts of alkaloid do not appear to be greater than that reported for untransformed lines of *C. ledgeriana* [62,63]. Quinine and quinidine are of commercial importance since quinine is used as an antimalarial drug and as a bitter flavoring in the soft-drink industry, while quinidine is used to treat cardiac arrhythmias.

Camptothecin obtained from cell cultures of the Chinese species *Camp-*

totheca acuminata (Nyssaceae) (Table II) [66–68] is another example of a tryptophan-derived quinoline alkaloid which is biosynthesized via indole intermediates. This alkaloid has marked antitumor activity, but its clinical use was abandoned because of its high toxicity.

B. Tyrosine-Derived Alkaloids

The production of isoquinoline alkaloids by plant cell cultures has been the subject of a number of comprehensive reviews [69,70]. To date some 60 isoquinoline alkaloids have been identified from plant cell cultures (Table III), representing 11 of the major structural types: benzylisoquinolines, bisbenzylisoquinolines, proaporphines, aporphines, protoberberines, proto-pines, benzophenanthridines, phthalideisoquinolines, morphinans, aristolac-tams, and *Cephalotaxus* alkaloids. Some 35 species from 20 genera have yielded isoquinoline alkaloids, and this is a greater number of plants than those which produce indole alkaloids. The families from which these genera are derived include Berberidaceae, Cephalotaxaceae, Fumariaceae, Meni-spermaceae, Papaveraceae, and Ranunculaceae.

Some of the major yields in the production of secondary metabolites from cell cultures have been achieved with isoquinoline alkaloids. In 1981, a yield of 2.7 g liter^{-1} of jatrorrhizine from *Berberis stolonifera* cultures [82] was reported, and a recent study has revealed that *B. wilsoniae* grown in 20-liter airlift bioreactors produces more than 3 g liter^{-1} of alkaloid [81]. High berberine-producing cultures have been derived from *Coptis japonica*, with selected lines reaching 13.2% berberine on a dry weight basis (1.39 g liter^{-1}) [84].

Rueffer, in her comprehensive review of isoquinoline alkaloids in plant cell cultures, noted that the protoberberine group of alkaloids shows the greatest variety in its aromatic substitution pattern [69]. To date some 17 protoberberine alkaloids have been found in cultures initiated from some 20 species (Table III). Three species which produce an extremely wide range of protoberberine alkaloids in cell culture are *Coptis japonica* (9 alkaloids), *Nandina domestica* (10 alkaloids), and *Thalictrum minus* (11 alkaloids). In addition, cultures of these three species also produce the aporphine alkaloid magnoflorine. A number of species from the genera *Chasmanthera, Cheli-donium, Corydalis, Dicentra, Eschscholtzia,* and *Tinospora* have been shown to produce up to three different structural types of isoquinolines in cell cultures, while *Stephania, Papaver,* and *Fumaria* species produce an even wider range (Table III). *Stephania* cell cultures produce aporphine, bisbenzy-lisoquinoline, and protoberberine alkaloids. So far, *Stephania* is the only genus reported to produce aristolactam-type alkaloids (cepharanones I and II) in cell cultures [69,104].

TABLE III. Tyrosine-Derived Alkaloids From Plant Cell Cultures

Structural type	Alkaloid	Source [reference]
Isoquinolines		
Benzylisoquinolines	Reticuline	*Chasmanthera dependens*
	Methylcoclaurine, reticuline	[69] *Fumaria capreolata* [71]
Bisbenzylisoquinolines	Aromoline, berbamine	*Stephania cepharantha* [72]
Proaporphines	Stepharine	*Tinospora caffra* [69] *T. cordifolia*
Aporphines	Cepharadione A, cepharadione B, liriodenine, lysicamine, norcephradione	*Stephania cepharantha* [73]
	Isoboldine Magnoflorine	*Fumaria capreolata* [71] *Chasmanthera dependens* [69] *Coptis japonica* [74] *Corydalis incisa* [75] *Corydalis pallida* [75] *Dicentra peregrina* [75] *Dioscoreophyllum cumminsi* [76] *Eschscholtzia californica* [75] *Fumaria capreolata* [71] *Mahonia japonica* [77] *Nandina domestica* [77] *Papaver bracteatum* [75] *P. orientale* [75] *P. rhoeas* [75, 92]

TABLE III. Tyrosine-Derived Alkaloids From Plant Cell Cultures (continued)

Structural type	Alkaloid	Source [reference]
		P. setigerum [75]
		P. somniferum [75]
		Stephania japonica [69]
		Thalictrum minus [74]
		Tinospora caffra [69]
	Stephanine	*Stephania glabra* [79]
Protoberberines	Berberine	*Argemone mexicana* [80]
	Berberine, columbamine, jatrorrhizine	*Berberis stolonifera* [75,82]
	Berberine	*B. vulgaris* [69]
	Berberine, columbamine, jatrorrhizine, palmatine	*B. wilsoniae* [81]
	Dehydrocorydalmine, jatrorrhizine, palmatine	*Chasmanthera dependens* [69]
	Coptisine	*Chelidonium majus* [69]
	Berberine, columbamine, coptisine, epiberberine, groenlandicine, jatrorrhizine, palmatine, thalidastine, berberastine	*Coptis japonica* [74,77,83–85]
	Jatrorrhizine, palmatine	*Dioscoreophyllum cumminsii* [76]
	Coptisine, dehydrocheilanthifoline, scoulerine	*Fumaria capreolata* [71]
	Berberine, columbamine, coptisine, jatrorrhizine, palmatine	*Mahonia japonica* [77]
	Palmatine	*Menispermum canadense* [69]
	Berberine, berberastine, columbamine, coptisine, epiberberine, groenlandicine, jatrorrhizine, palmatine, thalifendine, thalidastine	*Nandina domestica* [77]
	Orientalidine, stylopine	*Papaver bracteatum* [86,91]
	Orientalidine	*P. setigerum* [86]
		P. somniferum
	Berberine, palmatine	*Phellodendron amurense* [69]
	Palmatine, jatrorrhizine	*Stephania japonica* [69]
	Berberine, columbamine, coptisine, desoxythalistadine, epiberberine, jatrorrhizine, palmatine, thalifendine, thalidastine, berberastine, desoxythalidastine	*Thalictrum minus* [74,77,78,85]

TABLE III. Tyrosine-Derived Alkaloids From Plant Cell Cultures (continued)

Structural type	Alkaloid	Source [reference]
Protopines	Columbamine, jatrorrhi-zine, palmatine	*Tinospora caffra* [69]
	Jatrorrhizine, palmatine	*T. crispa* [69]
	Protopine	*Chelidonium japonica* [75]
	Allocryptopine, protopine	*C. majus* [69]
	Protopine	*Corydalis incisa* [75]
		C. ophiocarpa [87]
		C. pallida [75]
		Dicentra peregrina [75]
		Eschscholtzia californica [75]
		Fumuria capreolata [71]
		Macleaya cordata [75]
		Papaver bracteatum [75,91]
		P. nudicale [86]
		P. orientale [75]
		P. rhoeas [75]
		P. setigerum [75,86]
	Allocryptopine, protopine	*Macleaya microcarpa* [88]
	Cryptopine, protopine	*Papaver somniferum* [75,86,89,90,92]
Benzophen-anthridines	Dihydrosanguinarine, norsanguinarine, oxosanguinarine	*Chelidonium japonica* [75]
	Chelidonine, sanguinarine	*Chelidonium majus* [69]
	Sanguinarine	*Corydalis ophiocarpa* [87]
	Norsanguinarine	*C. incisa* [75]
	Oxysanguinarine	*C. pallida* [75]
	Dihydrosanguinarine, sanguinarine	*Dicentra peregrina* [75]
	Chelirubine, chelerythrine, dihydrochelerythrine, dihydrochelirubine, dihydromacarpine, dihydrosanguinarine, norsanguinarine, macarpine, oxosan-guinarine, sanguinarine	*Eschscholtzia californica* [75,93]
	Sanguinarine	*Fumaria capreolata* [71]
	Chelirubine, dihydro-sanguinarine, norsanguinarine, oxosanguinarine, sanguinarine	*Macleaya cordata* [75]
	Sanguinarine	*Macleaya microcarpa* [88]

TABLE III. Tyrosine-Derived Alkaloids From Plant Cell Cultures (continued)

Structural type	Alkaloid	Source [reference]
		Papaver bracteatum [86,94]
		P. nudicale
		P. setigerum
		P. somniferum
	Dihydrosanguinarine	*P. bracteatum* [75,92]
	Norsanguinarine	*P. orientale* [75,92]
	Oxosanguinarine	*P. rhoeas* [75,92]
	Sanguinarine	*P. setigerum* [75,92]
		P. somniferum [75,92]
	Chelirubine	*P. bracteatum* [75]
	Acetonyldihydrosanguinarine	*P. somniferum* [92]
Phthalideisoquin-	Narcotine	*Papaver rhoeas* [95]
olines	Narceine, narcotine	*P. somniferum* [96]

Morphinans	Pallidine	*Fumaria capreolata* [71]
	Thebaine	*Papaver bracteatum* [94,105, 106[a],107]
	Codeine, thebaine, morphine	*P. rhoeas* [95]
	Codeine	*P. setigerum* [69]
	Codeine, morphine, thebaine, isothebaine, papaverine	*P. somniferum* [86,96–103,108]
Aristolactams	Cepharanone I and II	*Stephania cepharantha* [104]

Cephalotaxus alkaloids	Harringtonine, homoharringonine, isoharringtonine, homodeoxyharringtonine cephalotaxine	*Cephalotaxus harringtonia* [109–111]

---

[a]Root cultures.

Interestingly, *Papaver* species in cell culture also produce an extensive range of isoquinoline alkaloids including aporphine, protoberberine, protopine, benzophenanthridine, phthalideisoquinoline, and morphinan types. Many of these alkaloids are not typical of the parent plants. Despite the obvious capability of *Papaver* cultures to produce isoquinoline alkaloids, codeine and morphine, which are the most important from a pharmaceutical viewpoint, have proved extremely difficult to produce.

The most diverse range of isoquinoline-type alkaloids ever reported from a cell suspension culture has been obtained from *Fumaria capreolata* [71]. Ten isoquinoline alkaloids, representing six structural classes (Table III), have been found in this one species, viz, benzylisoquinoline (methylcoclaurine, reticuline), aporphine (isoboldine, magnoflorine), protoberberine (coptisine, dehydrocheilanthifoline, scoulerine), protopine (protopine), benzophenanthridine (sanguinarine), and morphinan (pallidine).

In addition to the extensive range of known isoquinoline alkaloids which have been identified from plant cell cultures, several novel compounds have also been found. The novel oxoaporphine norcepharadione has been isolated from callus tissue of *Stephania cepharantha* [73], and norsanguinarine was located as a natural product for the first time from callus cultures of *Papaver somniferum* [92]. Plant cell cultures have also been shown to produce isoquinoline alkaloids which are not typical of the parent plant and in some instances new to a particular genus. Many of the alkaloids from *Papaver* cell cultures, e.g., stylopine and protopine [91], have not been identified in the corresponding parent plant. Aromoline, a bisbenzylisoquinoline, found in *Stephania* cultures, has not previously been identified from the genus [72].

C. Lysine-Derived Quinolizidine Alkaloids

Plants producing quinolizidine alkaloids occur in a number of families, but they are particularly abundant in the Leguminosae (Fabaceae). Ten species of the Leguminosae, comprising six genera, i.e., *Baptisia, Cytisus, Genista, Laburnum, Lupinus*, and *Sophora*, have been extensively investigated in cell culture by Wink et al. and Wink and Hartmann (Table IV) [112–116]. In general, the alkaloids produced by the cell cultures differed both quantitatively and qualitatively from the parent plant. Overall, the level of alkaloid produced by the cultures was several orders of magnitude lower than the parent species. Furthermore, the alkaloid pattern of the cell suspension cultures was much simpler than the whole plants and differed markedly in that the α-pyridone alkaloids, such as cytisine and N-methylcytisine, which predominate in the plant, could not be detected in the cell cultures. Twelve quinolizidine alkaloids were, however, identified in cell cultures of the ten species investigated [112]. Lupanine proved to be the

TABLE IV. Lysine-Derived Quinolizidine Alkaloids From Plant Cell Cultures

Structural type	Alkaloid	Source [reference]
Quinolizidines		
Tricyclics	11-allylcytisine, tinctorine	*Cytisus canariensis* [112]
Tetracyclics	Lupanine, 4-hydroxylupanine, 13-hydroxylupanine, 17-oxolupanine, 13-angeloyloxylupanine, 13-tigoyloxy-lupanine, 13-cis-cinnamoyloxy-lupanine, 13-trans-cinnamoyloxy-lupanine, sparteine, 17-oxosparteine, tetrahydrorhombifoline	*Lupinus polyphyllus* [112–115]
	Lupanine, 4-hydroxylupanine, 17-oxolupanine, sparteine, 17-oxosparteine, tetrahydrorhombifoline	*Lupinus hartwegii* [112,115]
	Lupanine, sparteine, tetrahydro-rhombifoline	*L. luteus* [112]
	Lupanine, sparteine	*Cytisus scoparius* [112]
	Lupanine, tetrahydrorhombifoline	*C. purpureus* [112]
		Genista pilosa [112]
	Lupanine	*Baptisia australis* [112,116]
		Cytisus canariensis [112]
		Laburnum alpinum [112]
		Sophora japonica [112]

major alkaloid in cultures of all ten species and was accompanied by sparteine, tetrahydrorhombifoline, 17-oxosparteine, 13-hydroxylupanine, 4-hydroxylupanine, 17-oxolupanine, and 13-hydroxylupanine esters as minor alkaloids in some species (Table IV).

D. Ornithine-Derived Tropane Alkaloids

Tropane alkaloids, such as atropine((±)-hyoscyamine), (−)-hyoscy-amine, and scopolamine, are medicinally important secondary plant products because of their action on the autonomic nervous system. These alkaloids are synthesized mainly in the roots of some Solanaceous plants, and studies with in vitro cultures indicate that the production of significant levels of tropane alkaloids depends on the degree of organogenesis of the culture. Tropane alkaloids have been reported from undifferentiated callus cultures of various genera of the Solanaceae; however, the majority of reports involve highly differentiated cultures, in particular, root cultures (Table V).

TABLE V. Ornithine-Derived Tropane Alkaloids From Plant Tissue Cultures

Structural type	Alkaloid	Source [reference]
Tropanes	Atropine ((±)-hyoscyamine)	*Atropa belladonna* [117–118[a]]
	Scopolamine (hyoscine)	*Datura stramonium* [119]
RN	Atropine, scopolamine, tropine	*D. innoxia* [120]
	Hyoscyamine, scopolamine	*Duboisia hopwoodii* [121[a]]
		D. leichardtii [121[a],122]
		D. myoporoides [121[a],123[a]]
	Atropine, valtropine	*D. myoporoides* [123[a],124,125]
	Hyoscyamine, littorine,	*Hyoscyamus albus* [126[a]]
	6-β-hydroxyhyoscyamine	
	6-β-hydroxyhyoscyamine, littorine	*H. gyorffi* [126[a]]
		H. pusillus
	6-β-hydroxyhyoscyamine,	*H. muticus* [126[a],127,128]
	scopolamine, hyoscyamine	
	Hyoscyamine, scopolamine, 6-β-	*H. niger* [126[a],129[a],130,131[a]]
	hydroxyhoscyamine, cuscohygrine	
	6-β-hydroxyhyoscyamine	*H. bohemicus* [126[a]]
		H. canariensis
	Hyoscyamine, scopolamine	*Scopolia acutangula* [132,133]
	Hyoscyamine, scopolamine,	*S. parviflora* [134]
	apoatropine	

[a]Root cultures.

Recent studies with *Duboisia* have shown that, in general, callus cultures produce only low levels of the desired tropanes, in contrast to differentiated roots obtained from callus, which produce the range of alkaloids typical of the parent plant [121–123]. Shoot cultures, regenerated from these *Duboisia* species, on the other hand, failed to produce tropane alkaloids; and these results correlate well with the proposed association of tropane alkaloid production with root organogenesis.

Yamada and co-workers have initiated callus cultures of *Atropa belladonna*, *Datura stramonium*, and *Hyoscyamus niger*; but only *H. niger* was found to contain significant levels of alkaloids [126,129–131]. The main alkaloid, hyoscyamine, could be increased to a level comparable with the whole plant by means of cell selection and optimization of cultural conditions. The scopolamine content of the callus, however, remained low, being about one-tenth of the parent plant, but this was increased significantly by inducing root formation in the cultures, indicating that differentiation is required for scopolamine biosynthesis in *H. niger* [126,129–131].

Similarly, studies with *Scopolia parviflora* cultures have shown a close relationship between organization of the tissue and alkaloid formation [134].

TABLE VI. i) Nicotinic-Acid-Derived Pyridines and ii) Purine Alkaloids From Plant Cell Cultures

Structural type	Alkaloid	Source [reference]
Pyridines	Nicotine	*Duboisia leichardtii* [121ᵃ,122ᵃ]
		D. hopwoodii
	Nicotine, anabasine, nornicotine	*D. myoporoides* [121ᵃ,123ᵃ]
	Nicotine, nornicotine, anabasine, anatabine	*Nicotiana rustica* [135–137,138ᵇ]
	Nicotine, nornicotine, myosine, anabasine, anatabine, nicotelline, anatelline	*N. tabacum* [137,139–141,142ᵇ]
Purines	Caffeine, theobromine	*Coffea arabica* [143–145]
	Caffeine	*Camellia sinensis* [146,147]

ᵃRoot cultures.
ᵇCultures transformed with *Agrobacterium rhizogenes*.

Callus and suspension cultures with no obvious microscopical differentiation produced only small amounts of tropane alkaloids, whereas roots initiated from callus cultures contained the normal pattern of alkaloids.

E. i) Nicotinic-Acid-Derived and ii) Purine Alkaloids

Nicotine-type alkaloids have been produced by cell cultures of *Duboisia* and *Nicotiana* species (Table VI). Cultured roots of *Duboisia* species have yielded nicotine and nornicotine [121–123], and traces of anabasine have been reported from *D. myoporoides* callus [123]. *Nicotiana tabacum* callus has yielded 3.75% total alkaloid calculated on a dry weight of cells basis and the major alkaloid was nornicotine [136]. "Hairy root" cultures of *N. rustica* have been developed by transformation with *Agrobacterium rhizogenes*, and nicotine was released into the medium at a maximum concentration of 10 mg liter^{-1} [138,142]. It has been suggested that "hairy root" cultures have considerable potential for the production of valuable secondary products when continuous fermentation techniques are applied [138]. Suspension cultures of *N. tabacum* produce other tobacco alkaloids, including anatabine, anabasine, myosmine, anatelline, and nicotelline [141].

No recent publications on the production of the purine alkaloids by cell cultures (Table VI) were noted during the preparation of this chapter.

TABLE VII. Anthranilic Acid Derived Alkaloids From Plant Cell Cultures

Structural type	Alkaloid	Source [reference]
Furoquinolines	Balfourodine, platydesmine	*Choisya ternata* [148]
	Platydesmine, ribaline, rutaline, skimmianine, kokusaginine, 6-methoxydictamnine	*Ruta graveolens* [149–152]
Acridones	1-hydroxyrutacridone epoxide	*Ruta bracteosa* [153]
		R. chalepensis
		R. corsica
		R. macrophylla
	Hydroxyrutacridone epoxide, gravacridonol, rutagravin, hydroxy-3-methoxy-N-methyl-acridone, hyroxy-N-methyl-acridone, rutacridone, rutacridone, epoxide	*R. graveolens* [154–156]
Quinolones	Edulinine	*R. graveolens* [149–151]

F. Anthranilic-Acid-Derived Alkaloids

Acridone, quinolone, and furoquinoline alkaloids are typical secondary metabolites of the family Rutaceae. Only two genera of this family have been studied in cell culture, viz., *Ruta* and *Choisya* (Table VII). Callus and cell suspension cultures of *Ruta graveolens* have been shown to produce a diverse range of alkaloids, including furoquinolines (six alkaloids), acridones (seven alkaloids), and one quinolone (edulinine). Two new alkaloids, termed furacridones, have been isolated for the first time from callus cultures of *R. graveolens*, they are 1-hydroxyrutacridone epoxide and rutagravin, which represents a new pentacyclic system of acridone alkaloids with a difuranoid moiety [156] (Fig. 3). Callus and suspension cultures of other *Ruta* species— *R. macrophylla, R. corsica, R. chalepensis*, and *R. bracteosa*—also produce rutacridone epoxide and hydroxyrutacridone epoxide [153]. Two dihydrofuroquinoline alkaloids, platydesmine and balfourodine, have been identified from cultures of *Choisya ternata* [148].

G. Acetate-Derived Alkaloids

Eight steroidal alkaloids have been reported from cultures of ten *Solanum* species, and tomatine has been obtained from cultures of *Lycopersicum esculentum* (Table VIII) [157–173]. Total alkaloid concentrations of 0.1 mg

rutagravin   1-hydroxyrutacridone epoxide

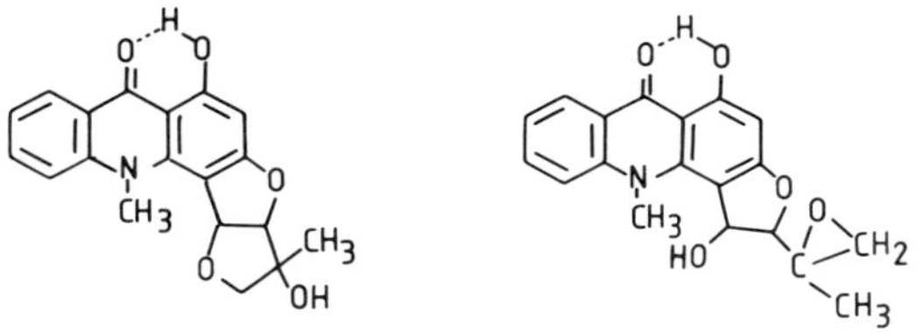

Fig. 3. Rutagravin and 1-hydroxyrutacridone epoxide.

TABLE VIII. Acetate-Derived Alkaloids From Plant Cell Cultures

Structural type	Alkaloid	Source [reference]
Steroidal	Tomatine	*Lycopersicum esculentum* [157]
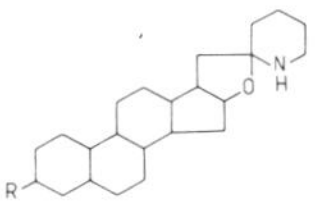	Solasodine, solamargine	*Solanum acculeatissimum* [159]
	Dehydrocommersonine	*S. chacoense* [160ª]
	Solasodine, soladulcidine	*S. dulcamara* [161, 162]
	Solasodine	*S. jasminoides* [163]
	Solasodine, solasonine, solanidine	*S. khasianum* [164,165ª,166]
	Solasodine	*S. laciniatum* [167–169]
	Solasodine	*S. nigrum* [170]
	Solanine, chaconine	*S. tuberosum* [160ª]
	Solasodine	*S. verbascifolium* [171]
	Solasodine, solasonine	*S. xanthocarpum* [172,173]

ªRoot cultures.

g^{-1} dry weight of green suspension cultures of *S. dulcamara* have been obtained in which soladulcidine and solasidine were the major alkaloids [161]. Factors affecting the production of solasidine in cultures of *S. lacinatum* [168,169] and *S. nigrum* [170] have been investigated.

II. STUDIES ON ALKALOID BIOSYNTHESIS USING PLANT CELL CULTURES

Major research efforts are being directed towards an understanding of the biosynthesis of indole and isoquinoline alkaloids, and this section deals with research papers which have been published since the previous review by Anderson et al. [180].

Cell suspension cultures of *Ailanthus altissima* have been fed with L-[methylene ^{14}C]-tryptophan and [sidechain 2 ^{14}C]-tryptamine which were rapidly taken up and incorporated into canthin-6-one, 1-hydroxy- and

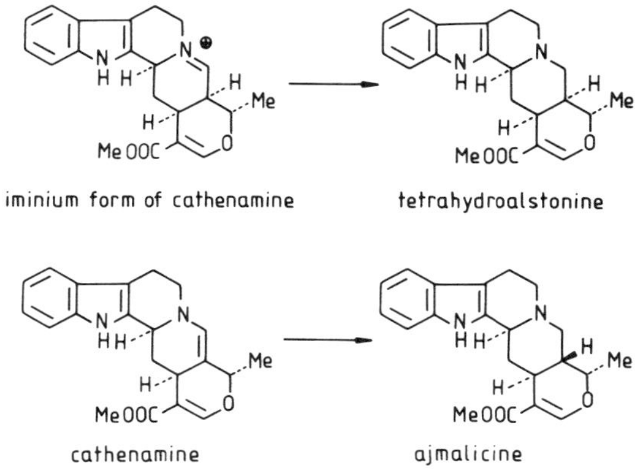

Fig. 4. Formation of tetrahydroalstonine and ajmalicine from cathenamine.

1-methoxy-canthin-6-one, [181,182]. L-[methyl ^{14}C]-methionine feeding of the same cultures has resulted only in radiolabeled 1-methoxycanthin-6-one, and no label was observed in the other two alkaloids [182].

Alkaloid yields in *Catharanthus roseus* cell cultures are apparently not affected by the levels of activity of tryptophan decarboxylase, which is one of the first committed enzymes of indole alkaloid biosynthesis [183]. Furthermore, no correlation has been observed between tryptophan decarboxylase activity and the accumulation of the corynanthean alkaloids ajmalicine and serpentine [184], and it has been suggested that the iridoid pathway is the limiting factor for indole alkaloid production in *C. roseus* cell cultures.

An authoritative article has reviewed the enzymatic biosynthesis of ajmalicine-type alkaloids and the utilization of plant cultures in such studies [185]. More recent studies have been concerned with the purification and characterization of the enzymes involved [186]. Two distinct cathenamine reductases have been identified; one reduces the iminium form of cathenamine to give tetrahydroalstonine, and the other enzyme reduces cathenamine to give ajmalicine (Fig. 4). The tetrahydroalstonine synthase utilizes NADPH and has been identified in cultures of *C. roseus, C. ovalis, Rhazya stricta*, and *Vinca herbacea*.

The enzymes involved in the late stages of vindoline biosynthesis continue to be of current interest, and although plant cell cultures have proved to be invaluable sources of enzymes, resort in some instances has to be made to whole plants. The specific enzyme acetyl-Co A: 17-O-deacetylvindoline

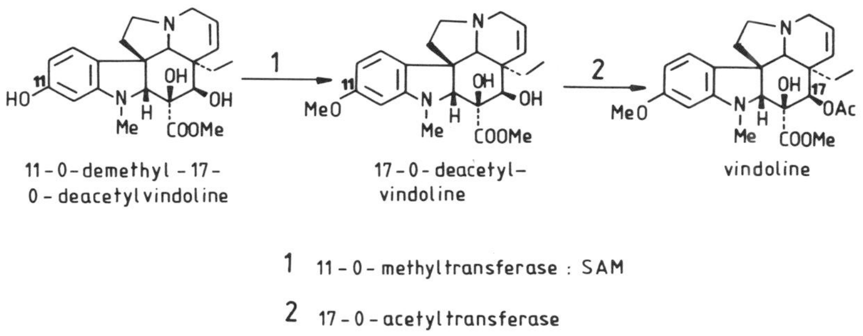

11-0-demethyl-17-
0-deacetylvindoline

17-0-deacetyl-
vindoline

vindoline

1 11-0-methyltransferase : SAM

2 17-0-acetyltransferase

Fig. 5. Late stages in the biosynthesis of vindoline.

17-O-acetyltransferase, which catalyses the reverse reaction of hydrolysis of the 17-O-acetyl group, has been isolated from differentiated plants of *C. roseus* but was not detectable in cell suspension cultures, which contrasts with the results of other workers [187]. Similarly, S-adenosyl-L-methionine:11-O-demethyl-17-O-deacetylvindoline 11-O-methyltransferase (Fig. 5), which has been obtained from the whole plant, has not yet been detected in cell cultures of *C. roseus* [188]. Studies with cell-free extracts of cell suspensions from *C. roseus* have demonstrated enzymic activity which can couple the indole alkaloids, vindoline and catharanthine, to dimeric alkaloids [189]. The major product obtained was 3′, 4′ − anhydrovinblastine together with other dimeric alkaloids, including leurosine, catharine, vinamidine, and 3(R)-hydroxy-vinamidine.

A number of ^{2}H- and ^{13}C-labeled monoterpenes have been fed to *Rauwolfia serpentina* suspension cultures in order to study the early stages of ajmaline and vomilenine biosynthesis [190]. NMR studies of the isolated alkaloids were used to determine the location of the incorporated ^{2}H and ^{13}C atoms. The results clearly demonstrated that iridodial, 10-hydroxygeraniol, and 10-hydroxynerol were incorporated into these two alkaloids via 10-oxogeranial and 10-oxoneral and therefore are biosynthesized in *R. serpentina* cell cultures via the same process as secologanin and vindoline in whole plants of *Lonicera* and *Catharanthus*.

Vellosimine reductase has been isolated from cell suspension cultures of *R. serpentina*, two other species of *Rauwolfia* and *C. roseus* [191]. The enzyme catalyses the NADPH-dependent conversion of vellosimine into 10-deoxysarpagine (Fig. 6). Vinorine synthase, which is a key enzyme in the formation of ajmaline, has been isolated from cell suspension cultures of *R. serpentina* [192]. Vinorine is the precursor of alkaloids in the final stages of

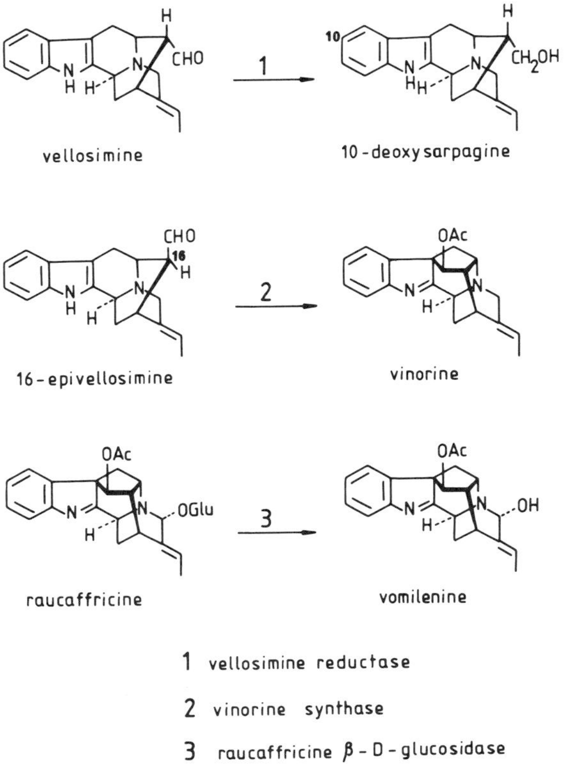

Fig. 6. Late stages in the biosynthesis of sarpagine- and ajmaline-type alkaloids.

the ajmaline pathway, vomilenine (21-hydroxyvinorine), 17-O-acetylnoraj-maline, and 17-O-acetylajmaline. Raucaffricine, also isolated from *R. serpentina* cell cultures in yields of 1.2 g liter^{-1} of medium, was originally characterized as vomilenine-galactoside but its structure has been revised to vomilenine-β-D-glucoside [176,193]. The highly specific raucaffricine-β-D-glucosidase has been isolated from the same cell suspension cultures and shown to catalyze the formation of vomilenine from raucaffricine (Fig. 6).

The radiolabels from L-[methylene ^{14}C]-tryptophan, [sidechain 2 ^{14}C]-tryptamine, and [C-5 ^{3}H]-secolaganin have been shown to be incorporated into quinine and quinidine when fed to root organ cultures of *Cinchona ledgeriana* [194,195]. A high activity of tryptophan decarboxylase in cell cultures of *C. succirubra* is said to be a prerequisite for alkaloid produc-

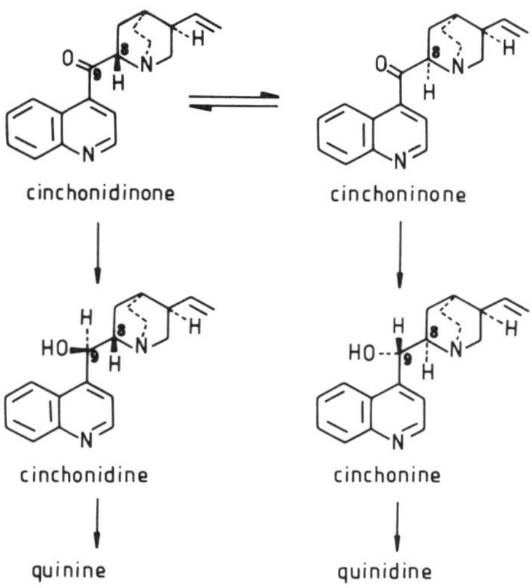

Fig. 7. The biosynthesis of quinine and quinidine from cinchoninone.

tion [65], and it has been proposed that the activity of this enzyme is potentially a rate-limiting step for alkaloid production in *C. ledgeriana* cultures [196]. Two isoenzymes of cinchoninone:NADPH oxidoreductase have been isolated [197]. Isoenzyme I acts specifically on cinchoninone in the forward direction and on cinchonidine, cinchonine, cupreine, and cupreidine in the reverse direction. Isoenzyme II has broad specificity, acting on all of the quinoline alkaloids of *Cinchona*. Enzymological evidence indicates that the 8-R ketone cinchoninone is a key intermediate in the latter stages of the biosynthesis of quinine and quinidine (Fig. 7).

Biosynthetic studies utilizing plant cell cultures which produce isoquinoline alkaloids have also proceeded at the enzyme level [10,198]. The use of cell free extracts of *Berberis* cultures have enabled the isolation of eight enzymes which are responsible for the biosynthesis of berberine (Fig. 8). Recent publications have described the purification and characterization of several of these enzymes. *B. vulgaris* cell suspension cultures have yielded two N-methyltransferases (NMT-I and NMT-II). NMT-I, obtained from *B. wilsoniae*, is specific for tetrahydrobenzylisoquinoline alkaloids and utilizes S-adenosyl-L-methionine (SAM) as the methyl donor. This enzyme has been used to prepare ^{14}C and ^{3}H-labeled N-methylbenzyltetrahydroisoquinolines [199]. The berberine bridge-forming enzyme, purified from *B. beaniana* cell cultures, is also present in cell cultures from other species of the Berberida-

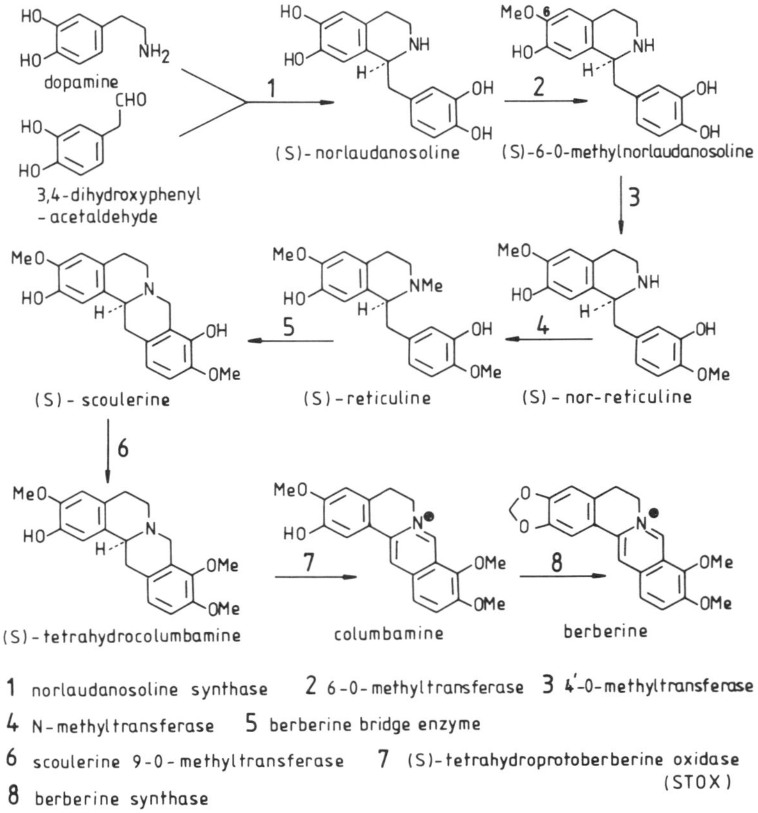

Fig. 8. The biosynthesis of berberine.

ceae and from species of Ranunculaceae, Menispermaceae, Papaveraceae, and Fumariaceae [200]. (S)-Adenosyl-L-methionine:(S)-scoulerine 9-O-methyltransferase, a highly stereo- and regiospecific enzyme of tetrahydro-protoberberine synthesis [201], and (S)-adenosyl-L-methionine:colum-bamine-O-methyltransferase [202] have been obtained from *B. wilsoniae* cell cultures. Columbamine is converted to palmatine by a specific methyltrans-ferase which acts only on this quaternary alkaloid [203] and has been obtained from two species of *Berberis* cell cultures. (S)-Tetrahydroberberine has been converted to berberine by a crude enzyme preparation from cell cultures of *Coptis japonica*, indicating that there is more than one route to the biosynthesis of berberine [204].

Two of the enzymes of berberine biosynthesis, the berberine bridge enzyme and (S)-tetrahydroprotoberberine oxidase (STOX) are exclusively located in highly specific vesicles which are found not only in *Berberis* cell cultures but also in cell cultures obtained from species of Annonaceae, Menispermaceae, and Ranunculaceae. It would appear, therefore, that (S)-scoulerine, the product of the berberine bridge enzyme, will have to leave these vesicles in order to be converted into (S)-tetrahydrocolumbamine within the cytoplasm, and that this alkaloid will enter the vesicles to act as a substrate for STOX (Fig. 8) [205].

The first enzyme in the biosynthesis of quinolizidine alkaloids is lysine decarboxylase, which converts lysine to cadaverine; the sequences which lead to a series of quinolizidine alkaloids [180] and the metabolism of these alkaloids in plants and in cell cultures [206] have been reviewed. Several quinolizidine alkaloid-yielding genera produce only low yields when grown in cell cultures; e.g., *Lupinus* species may accumulate as much as 5 mg alkaloid g^{-1} fresh weight, whereas cell cultures have levels of 0.01–1 $\mu g\ g^{-1}$ fresh weight.

Cell-free extracts of *Hyoscyamus niger* can convert hyoscyamine to 6-β-hydroxyhyoscyamine and responsible hydrolase has been isolated and characterized [207]. It has been argued, therefore, that hyoscyamine is converted into its β-6,7-epoxide, scopolamine, via the 6-β-hydroxy derivative and not via 6,7-dehydrohyoscyamine as previously postulated. Increased pyridine alkaloid levels in *Nicotiana tabacum* callus have been correlated with arginine decarboxylase activity, and it has been suggested that this enzyme plays a role in the overall biosynthesis of *Nicotiana* alkaloids [135].

Cell-free extracts of *Ruta graveolens* cell suspension cultures activate anthranilic acid, which is the second pathway-specific step in acridone alkaloid biosynthesis [208] and they synthesize 1,3-dihydroxy-N-methyl-acridone from N-methylanthranilic acid and malonyl co-enzyme A [209] (Fig. 9). This latter finding represents the first time that a cell-free system has proved to be capable of synthesizing an acridone alkaloid.

The biogenesis of steroids, including steroidal alkaloids, has been reviewed; and mention has been made of the use of cell culture techniques for biogenetic investigations [158].

III. LOCALIZATION OF ALKALOIDS IN PLANT CELL CULTURES

Although it has been suggested that the lack of gene expression of specific enzymes of secondary metabolism is the main reason for cell cultures being unproductive, some importance must be attached to mechanisms which permit accumulation rather than metabolism of alkaloids. In higher plants,

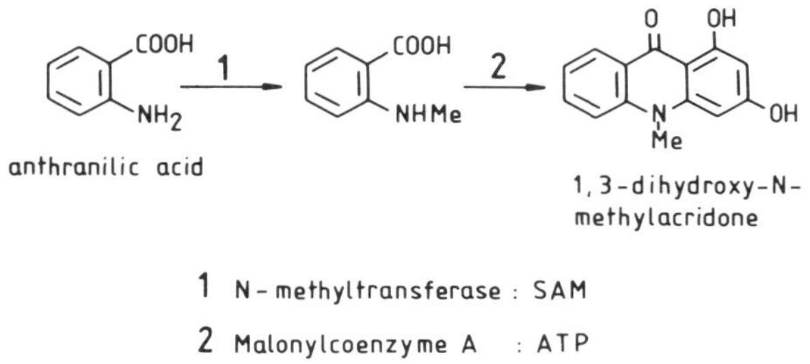

Fig. 9. The biosynthesis of acridone alkaloids.

specific tissues are frequently used for the accumulation of alkaloids; for example, the bark in *Cinchona* [57], the laticiferous system in *Papaver* [210], and the leaf epidermal and subepidermal layers in *Lupinus* species [211]. Other plants may accumulate alkaloids in nonspecific tissue with the development of idioblastic cells within the mesophyll, for example, the acridones in *Ruta graveolens* [212], the indoles in *Catharanthus roseus* [213], and isoquinolines in *Macleaya* species [214]. In every instance, it has been shown that the alkaloids are found in vacuoles within the cells. In some instances the required metabolic enzymes are absent from the cell cultures, but the cell vacuoles are able to accumulate efficiently the specific alkaloids of that species. For example, in cell cultures of *Senecio* species the alkaloids specifically accumulate as their N-oxides, which do not easily permeate membranes and are therefore ideal storage forms [215]. In *Lupinus* cultures, on the other hand, the enzymes required for alkaloid formation are present but alkaloids do not accumulate in large amounts. In this instance it would seem that a lack of adequate storage facilities allows for rapid further metabolism [206]. *Papaver*, in tissue culture, produces low levels of nonmorphinan alkaloids [216], but the formation of the morphinans appears to be correlated with the development of some differentiation within the cell culture [94,103,217,218]. It has also been suggested that a correlation exists between the appearance of the morphinan alkaloids and the onset of the development of laticiferous cells in germinating seedlings of *P. bracteatum* [219]. Whether it is the lack of storage capacity or the absence of certain key enzymes that causes absence of morphinans in undifferentiated cell cultures has yet to be determined.

Other plants in cell cultures, such as *Ailanthus altissima* [16], *R. graveolens* [212], and *M. microcarpa* [214] are high yielding in alkaloids and have adequate biosynthetic activity as well as storage capacity. The cell cultures of *Ruta* and *Macleaya*, as with the whole plants, sequester alkaloids in the vacuoles of certain mesophyll cells and therefore accumulate alkaloids in undifferentiated and partially differentiated tissues.

The mechanisms which enable plant cell vacuoles to accumulate alkaloids have received some study, and the importance of their acidic nature as compared with the surrounding cytosol has been recognised. In studies with *C. roseus* cells, alkaloid-containing vacuoles were found to exhibit a vacuolar pH of 3.0, while "normal" cell vacuoles had a pH of 5.0, the cytosol having a pH in the order of 6.8 [220]. Since some *Catharanthus* alkaloids such as serpentine fluoresce, cell sorting using flow cytometry was possible, and isolated protoplasts produced "accumulator" clones (acidic vacuole, high alkaloid accumulation) or "excretor" clones (neutral vacuoles, low alkaloid accumulation) [221,222]. This process of selection also demonstrated a positive correlation between vacuolar acidity and alkaloid accumulation [221,222].

The laticiferous systems of *Chelidonium majus* [223] and *P. somniferum* [210,224] sequester 70–90% of the major isoquinoline alkaloids in the acidic vesicles of the latex, whereas in cell cultures, as exemplified by *C. roseus*, sequestered alkaloid appears to be of the order of 30% [225,226]. Cell vacuoles in both whole plants and cell suspensions have shown a high degree of specificity for the alkaloids endogenous to the particular plant [224,227].

Various models have been put forward for the mechanism of alkaloid uptake into the cell vacuole. Experiments with the vacuoles of *C. majus* suggested a process of simple diffusion with trapping within the vacuole as a result of protonation or binding to phenolic acids [228]. In experiments with vacuoles from *P. somniferum* latex, the minimal temperature sensitivity, the alkaloid specificity, and the pH dependence further suggested the involvement of a channel protein [224]. Researchers using plant cell cultures [220,229,230] have made similar proposals. In all cases it appeared that active transport across the vacuolar membrane was not involved; however, recently Deus-Neumann and Zenk [231] have produced convincing evidence that alkaloid accumulation in the cell vacuoles of *Fumaria capreolata* cell suspensions is the result of specific active transport since uptake into the vacuoles of (S)-reticuline and (S)-scoulerine was stimulated by ATP and Mg^{2+} and is severely reduced by the use of the ATPase inhibitor N,N′ − dicyclohexylcarbodiimide (DCCD). Such stimulation of alkaloid uptake has also been found in recent experiments with fractionated vesicles from *P. somniferum* latex [232]. Experiments with (S)-reticuline have shown that the alkaloid moves in and out of the cell vacuoles with complete efflux in 40 min, and it has been proposed that this process requires highly specific carriers which allow transport into, as well as out of, the vacuole and that therefore in this instance an ion-trap mechanism is not involved for the accumulation of alkaloids within the vacuole [227,231]. Other work [226] using *C. roseus* cells as isolated protoplasts has shown that ajmalicine and serpentine are naturally distributed between the cells and medium and it has been suggested that this equilibrium is basically a consequence of the alkaloid's ability to diffuse through membranes. It has also been suggested that the media of cell cultures may act as a lytic compartment [233]. The presence of alkaloids in the culture medium of plant cells has been reported many times for *C. roseus* cells [234-236], as well as for other plant species [90,110,237–239].

Equilibriation curves for endogenous ajmalicine and ^{14}C-ajmalicine added to the media, together with efflux experiments, suggested that two pools of ajmalicine are present inside the cells, 70–80% of which responds quickly to pH and dilution and can be considered as a freely diffusible pool. The remaining 20–30% of ajmalicine is diffusible only over a prolonged period, showing that a substantial portion of the alkaloid is firmly bound to molecules within the vacuole. In these experiments serpentine was more strongly bound within the cells than ajmalicine [226]. It has been concluded that 1) the ion-trapping model is basically valid for indole alkaloids, pH gradients being one of the forces driving alkaloid compartmentation and accumulation; and 2) other forces, probably related to the binding of alkaloids to cellular components, accentuate their accumulation within vacuoles. The distribution of alkaloids between cells and their media in a given system therefore depends on the relative importance of these two forces and will vary depending on the cell strain and culture conditions.

IV. CONCLUSIONS

Plant cell cultures offer possible alternative, commercial sources for secondary products, such as alkaloids, provided that the processes are economically viable. From the information collated in Tables II–VIII it is obvious that many plant cell cultures are capable of producing a wide range of alkaloids, including many economically important compounds. However, much remains to be accomplished in our understanding of the regulation of alkaloid production before commercially viable processes are available. At the present time, only the napthoquinone shikonin, used as both a dye and as a pharmaceutical, is produced on a commercial scale by Mitsui Petrochemical Industries, Japan. The cultures of *Lithospermum erythrorhizon* are reported to produce 1.5 g liter^{-1} shikonin derivatives, which represents a yield 15 times greater than that of the parent plant [240,241]. Other processes, including the production of berberine by *Coptis japonica* cultures, are reported to be in an advanced state of development [242].

The apparent lack of success in the production of the antileukemic, dimeric alkaloids vinblastine and vincristine by *Catharanthus roseus* cultures may be circumvented by utilizing a combination of whole plants and cell cultures [243]. It may be possible to produce the two monomers, catharanthine and vindoline, by cell culture techniques and subsequently join them either by chemical coupling (modified Polonovski reaction) or using a suitable enzyme system. Production of catharanthine by *C. roseus* cell cultures has already proved successful with yields five times greater than that of the parent plant; and the recent report of enzymic synthesis of 3′, 4′ − anhydrovinblastine from vindoline and catharanthine by cell-free extracts from *C. roseus* cultures illustrates the significant progress which has been made towards isolating the important coupling enzymes [189].

Plant cell cultures have made a significant contribution to our understanding of alkaloid biosynthesis and have highlighted the potential of cell cultures as sources of commercially useful enzymes. It is now possible to synthesize large quantities of strictosidine for biosynthetic studies and chemical syntheses, using immobilized strictosidine synthase, whereas previously this compound proved difficult to prepare and purify [244]. Similarly, the presence of raucaffricine-β-D-glucosidase in cultures of *Rauwolfia serpentina*, rich in raucaffricine, could readily lead to substantial production of vomilenine [42,193]. Vomilenine is a good starting material for the chemical synthesis of other indole alkaloids, and this process may prove to have useful commercial applications.

In addition to characterizing and purifying enzymes involved in the biosynthesis of secondary metabolites, studies investigating the localization and compartmentation of both the enzymes and final products will increase our understanding of how secondary metabolites are formed and stored in the cell culture.

Numerous factors in both the internal and external environment have been shown to affect alkaloid production in plant cell cultures, and it has been acknowledged for some time that components of the cell culture medium, in particular, phytohormones, carbon source, macro- and micronutrients, can have a profound effect on alkaloid yield [5,245,246]. Furthermore, the addition of appropriate precursors has, in some instances, proved successful in improving alkaloid productivity in cell cultures [181,194,247]. In addition to the media effects, the physical conditions of cultivation such as aeration, light, temperature, pH, and agitation are also important parameters to be considered in the production of alkaloids [5,245–249].

Over the last few years a considerable amount of work has focused on the use of elicitors to induce and, hopefully, enhance alkaloid production [246,250]. The elicitors used can either be crude pathogen preparations—for example, the addition of autoclaved *Verticillium dahliae* conidia or *Fusarium moniliforme* to *Papaver somniferum* cell suspensions caused a considerable increase in codeine and morphine formation [3]—or isolated preparations, such as chitosan, a fungal cell-wall component which has been shown to enhance the yields of acridone alkaloid epoxides in *Ruta graveolens* suspension cultures [251].

A relatively new technique in plant cell cultures is the immobilization of plant cells within matrices such as calcium aliginate beads or polyurethane reticulate foams. This technique can result in increased productivity by the cells and spontaneous release of the product into the culture medium, in some cases. If, however, the product is not released spontaneously, there is scope to induce the release by permeabilization of the plant cells [252–254].

Undoubtedly, cell selection has proved to be one of the major tools in improving the productivity of cell cultures. The simplest approaches have utilized selection based on visual methods, in particular, fluorescence microscopy for alkaloids such as serpentine and berberine [84,255]. Sensitive radio- and enzyme-immunoassay methods have been developed for some alkaloids including quinine [256], serpentine, and ajmalicine [257]. A major advance in cell selection methodology has been achieved by the development of flow cytometry techniques for plant cells. Flow cytometry has been used with considerable success to sort protoplasts derived from *Catharanthus roseus* cell suspensions [221,222], however, the full impact of this tool for enhancing productivity of cell cultures has still to be realized. Disregarding the high cost involved, the technique offers enormous advantages in terms of rapid automation and versatility. Despite extensive cell selection, Deus-Neumann and Zenk [255] have observed that indole alkaloid production in *C. roseus* cell suspension cultures is unstable. Serpentine production was monitored in six selected cell lines over a period of 8 yr, and rapid loss of productivity invariably occurred during the first few months of cultivation. Spontaneous recovery of the initial alkaloid production was not observed, leading the authors to suggest that clonal selection of high-yielding plant strains appears to favor inherent instability.

Before the potential of plant cell cultures as alternative sources of secondary metabolites can be fully realized it is necessary to overcome the economic constraints. To achieve this it is essential that the major problems of identifying suitable products, selecting high-yielding strains, enhancing and stabilizing product yield, and optimizing large-scale production receive further sustained investigation.

It is perhaps premature to envisage the impact of genetic manipulation on secondary metabolite production by cell cultures. Already the use of *Agrobacterium tumefaciens* and *A. rhizogenes* to transform plant tissue cultures have produced exciting developments, and the alkaloid yields reported with Solanaceous species, in particular, have given a new perspective to commercial reality [138]. Before progress can be made in this field a full knowledge of alkaloid enzymology will be essential; to date, the biosynthesis of only one alkaloid is fully understood. However, a recent report of the cloning and characterization of a lysine decarboxylase gene from the enterobacterium *Hafnia alvei* is encouraging. The ultimate aim of the work is to introduce the gene into the genome of plant cells, which produce lysine-derived alkaloids, in order to overcome the apparent low activity of plant lysine decarboxylase [258]. This approach could point the way for future strategies in the manipulation of plant cell cultures for secondary metabolite production.

ACKNOWLEDGMENTS

We are most grateful to Mrs. L. Lisgarten for computer-assisted literature retrieval, to Mrs. A. Cavanagh for drawing the figures, and to Mrs. J. Hallsworth for typing the manuscript.

REFERENCES

1. Farnsworth NR: The role of medicinal plants in drug development. In Krogsgaard-Larsen P, Brogger Christensen S, Kofud H (eds): "Natural Products and Drug Development." Copenhagen: Munksgaard, 1984, p 17.
2. Anderson LA, Phillipson JD, Roberts MF: Alkaloid production by plant cells. In Webb C, Marituma F (eds): "Process Possibilities for Plant and Animal Cell Cultures." Chichester: Ellis Horwood Ltd, 1987, p 172.
3. Heinstein PF: Future approaches to the formation of secondary natural products in plant cell suspension cultures. J Nat Prod 48:1, 1985.
4. Staba EJ: Milestones in plant tissue culture systems for the production of secondary products. J Nat Prod 48:203, 1985.
5. Rokem JS, Goldberg I: Secondary metabolites from plant cell suspension cultures: Methods for yield improvement. In Mizrahi A, van Wezel AL (eds): "Advances in Biotechnological Processes." New York: Alan R. Liss, Inc., 1985, Vol 4, p 241.
6. Fowler MW: Industrial applications of plant cell culture. In Yeoman MM (ed): "Plant Cell Culture Technology." Botanical Monographs 23. Oxford: Blackwell, 1986, p 202.
7. Whitaker RJ, Hashimoto T: Production of secondary metabolites. In Evans DA, Sharp WR, Ammirato PV, Yamada Y (eds): "Handbook of Plant Cell Culture." New York: Macmillan, 1986, p 264.
8. Zenk MH, El-Shagi H, Arens H, Stöckigt J, Weiler EW, Deus B: Formation of the indole alkaloids serpentine and ajmalicine in cell suspension cultures of *Catharanthus*. In Barz W, Reinhard E, Zenk MH (eds): "Plant Tissue Culture and Its Biotechnological Application." Berlin: Springer-Verlag, 1977, p 27.

9. Sato F, Yamada Y: High berberine-producing cultures of *Coptis japonica* cells. Phytochemistry 23:281, 1984.

10. Zenk MH: Enzymology of benzylisoquinoline alkaloid formation. In Phillipson JD, Roberts MF, Zenk MH (eds): "The Chemistry and Biology of Isoquinoline Alkaloids." Berlin: Springer-Verlag, 1985, p 240.

11. Nickell LG: Products. In Staba EJ (ed): "Plant Tissue Culture as a Source of Biochemicals." Boca Raton: CRC Press, 1980, p 235.

12. Staba EJ: Secondary metabolism and biotransformation. In Staba EJ (ed): "Plant Tissue Culture as a Source of Biochemicals." Boca Raton: CRC Press, 1980, p 59.

13. Dougall DK: Production of biologicals by plant cell cultures. In Petricciani JC, Hopps HE, Chapple PJ (eds): "Cell Substrates." New York: Plenum, 1979, p 135.

14. Sasse F, Hammer J, Berlin J: Fluorimetric and high-performance liquid chromatographic determination of harmane alkaloids in *Peganum harmala* cell cultures. J Chromatogr 194:234, 1980.

15. Sasse F, Heckenberg U, Berlin J: Accumulation of β-carboline alkaloids and serotonin by cell cultures of *Peganum harmala* L. Plant Physiol 69:400, 1982.

16. Anderson LA, Harris A, Phillipson JD: Production of cytotoxic canthin-6-one alkaloids by *Ailanthus altissima* plant cell cultures. J Nat Prod 46:374, 1983.

17. Anderson LA, Phillipson JD, Roberts MF: Aspects of alkaloid production by plant cell cultures. In Morris P, Scragg AH, Stafford A, Fowler MW (eds): "Secondary Metabolism in Plant Cell Cultures." Cambridge: Cambridge University Press, 1986, p 1.

18. Kisakurek MV, Leeuwenberg AJM, Hesse M: A chemotaxonomic investigation of the plant families of Apocynaceae, Loganiaceae and Rubiaceae by their indole alkaloid content. In Pelletier SW (ed): "Alkaloids: Chemical and Biological Perspectives." Volume 1. New York: John Wiley and Sons, 1983, p 211.

19. Cordell GA: "Introduction to Alkaloids." New York: John Wiley and Sons, 1981.

20. Carew DP: Tissue culture studies of *Catharanthus roseus*. In Taylor WI, Farnsworth NR (eds): "The Catharanthus Alkaloids." New York: Marcell Dekker Inc, 1975, p 193.

21. Kutney JP, Choi LSL, Kolodziejczyk P, Sleigh SK, Stuart KL, Worth BR, Kurz WGW, Chatson KB, Constabel F: Alkaloid production in *Catharanthus roseus* cell cultures: Isolation and characterisation of alkaloids from one cell line. Phytochemistry 19:2589, 1980.

22. Kurz WGW, Chatson KB, Constabel F, Kutney JP, Choi LSL, Kolodziejczyk P, Sleigh SK, Stuart KL, Worth BR: Alkaloid production in *Catharanthus roseus* cell cultures: Initial studies on cell lines and their alkaloid content. Phytochemistry 19:2583, 1980.

23. Scott AI, Mizukami H, Hirata T, Lee S-L: Formation of catharanthine, akuammicine and vindoline in *Catharanthus roseus* suspension cells. Phytochemistry 19:488, 1980.

24. Stöckigt J, Soll HJ: Indole alkaloids from cell suspension cultures of *Catharanthus roseus* and *C. ovalis*. Planta Med 40:22, 1980.

25. Kutney JP, Choi LSL, Kolodziejczyk P, Sleigh SK, Stuart KL, Worth BR, Kurz WGW, Chatson KB, Constabel F: Alkaloid production in *Catharanthus roseus* cell cultures. VII: Effect of parameter changes and catabolism studies on cell line PRL No. 953. Helv Chim Acta 64:1837, 1981.

26. Kurz WGW, Chatson KB, Constabel F, Kutney JP, Choi LSL, Kolodziejczyk P, Sleigh SK, Stuart KL, Worth BR: Alkaloid production in *Catharanthus roseus* cell cultures VIII. Planta Med 42:22, 1981.

27. Kohl W, Witte B, Hofle G: Alkaloids from *Catharanthus roseus* tissue cultures II. Z Naturforsch 36C:1153, 1981.

28. Kutney JP, Choi LSL, Kolodziejczyk P, Sleigh SK, Stuart KL, Worth BR: Alkaloid

production in *Catharanthus roseus* cell cultures. V: Alkaloids from the 176G, 299Y, 340Y and 951G Cell Lines. J Nat Prod 44:536, 1981.

29. Kurz WGW, Constabel F, Kutney JP: Biosynthesis and biotransformation of indole alkaloids in *Catharanthus roseus* cell cultures. In Fujiwara A (ed): "Plant Tissue Culture 1982." Tokyo: Japanese Association for Plant Tissue Culture, 1982, p 361.

30. Constabel F, Gaudet-LaPrairie P, Kurz WGW, Kutney JP: Alkaloid production in *Catharanthus roseus* cell cultures. XII: Biosynthetic capacity of callus from original explants and regenerated shoots. Plant Cell Rep 1:139, 1982.

31. Petiard V, Courtois D, Gueritte F, Langlois N, Mompon B: New alkaloids in plant tissue cultures. In Fujiwara A (ed): "Plant Tissue Culture 1982." Tokyo: Japanese Association for Plant Cell Culture, 1982, p 309.

32. Gueritte F, Langlois N, Petiard V: Metabolites secondaires isoles d'une culture de tissus de *Catharanthus roseus*. J Nat Prod 46:144, 1983.

33. Kutney JP, Choi LSL, Kolodziejczyk P, Sleigh SK, Stuart KL, Worth BR, Kurz WGW, Chatson KB, Constabel F: Alkaloid production in *Catharanthus roseus* cell cultures. III: Catharanthine and other alkaloids from the 200 GW cell line. Heterocycles 14:765, 1980.

34. Kohl W, Witte B, Sheldrick WS, Hofle G: Indolalkaloide aus *Catharanthus roseus*-zellculturen. IV: 16R-19,20-E-Isositsirikin,16R-19,20-Z-Isositsirikin und 21-Hydroxy-cyclolochnerin. Planta Med 50:242, 1984.

35. Mulder-Krieger Th, Verpoorte R, de Water A, van Gessel M, van Oeveren BCJA, Baerheim-Svendsen A: Identification of the alkaloids and anthraquinones in *Cinchona ledgeriana* callus cultures. Planta Med 46:19, 1982.

36. Mulder-Krieger Th, Verpoorte R, de Graaf YP, van der Kreek M, Baerheim-Svendsen A: The effects of plant growth regulators and culture conditions on the growth and alkaloid content of callus cultures of *Cinchona pubescens*. Planta Med 46:15, 1982.

37. Kouadio K, Chenieux JC, Rideau M, Viel C: Antitumor alkaloids in callus cultures of *Ochrosia elliptica*. J Nat Prod 47:872, 1984.

38. Kouadio K, Creche J, Chenieux JC, Rideau M, Viel C: Alkaloid production by *Ochrosia elliptica* cell suspension cultures. J Plant Physiol 118:277, 1985.

39. Pawelka K-H, Stöckigt J: Indole alkaloids from *Ochrosia elliptica* plant cell suspension cultures. Z Naturforsch 41C:381, 1986.

40. Pawelka K-H, Stöckigt J, Danieli B: Epchrosine—a new indole alkaloid isolated from plant cell cultures of *Ochrosia elliptica* Labill. Plant Cell Rep 5:147, 1986.

41. Stöckigt J, Pfitzner A, Firl J: Indole alkaloids from cell suspension cultures of *Rauwolfia serpentina* Benth. Plant Cell Rep 1:36, 1981:

42. Schubel H, Stöckigt J: RLCC-isolation of Raucaffricine from its most efficient source—cell suspension cultures of *Rauwolfia serpentina* Benth. Plant Cell Rep 3:72, 1984.

43. Yamamoto O, Yamada Y: Production of reserpine and its optimisation in cultured *Rauwolfia serpentina* Benth. cells. Plant Cell Rep 5:50, 1986.

44. Pawelka K-H, Stöckigt J: Major indole alkaloids produced in cell suspension cultures of *Rhazya stricta* Decaisne. Z Naturforsch 41C:385, 1986.

45. van der Heijden R, Brouwer RL, Verpoorte R, Wijnsma R, van Beek TA, Harkes PAA, Baerheim-Svendsen A: Indole alkaloids from a callus culture of *Tabernaemontana elegans*. Phytochemistry 25:843, 1986.

46. Stöckigt J, Pawelka K-H, Rother A, Deus B: Indole alkaloids from cell suspension cultures of *Stemmadenia tormentosa* and *Voacanga africana*. Z Naturforsch 37C:857, 1982.

47. Pawelka K-H, Stöckigt J: Indole alkaloids from cell suspension cultures of *Tabernaemontana divaricata* and *Tabernanthe iboga*. Plant Cell Rep 2:105, 1983.

48. Arens H, Borbe HO, Ulbrich B, Stöckigt J: Detection of pericine, a new CNS-active indole alkaloid from *Picralima nitida* cell suspension culture by opiate receptor binding studies. Planta Med 46:210, 1982.

49. Stöckigt J, Pawelka K-H, Tanahashi T, Danieli B, Hull WE: Voafrine A and voafrine B, new dimeric indole alkaloids from cell suspension cultures of *Voacanga africana* Stapf. Helv Chim Acta 66:2525, 1983.

50. Carew DP: Reserpine in a tissue culture of *Alstonia constricta* F Muell. Nature 207:89, 1965.

51. Harris AL, Nylund HB, Carew DP: Tissue culture studies of certain members of the Apocynaceae. Lloydia 27:322, 1964.

52. MacCarthy JJ, Ratcliffe D, Street HE: The effect of nutrient medium composition on the growth cycle of *Catharanthus roseus* (L.) G. Don. cells grown in batch culture. J. Exp Bot 31:1315, 1980.

53. Mulder-Krieger TH: "Alkaloids in Callus Cultures of *Cinchona ledgeriana* Moens and *Cinchona pubescens* Vahl." Ph.D. Thesis, University of Leiden, 1984.

54. Verpoorte R, Wijnsma R, Mulder-Krieger Th, Harkes PAA, Baerheim-Svendsen A: Plant cell and tissue culture of *Cinchona* species. In Neumann D (ed): "Primary and Secondary Metabolism of Plant Cell Cultures." Berlin: Springer-Verlag, 1985, p 196.

55. Wijnsma R, Verpoorte R: Quinoline alkaloids in cell and tissue cultures of *Cinchona* species. In Vasil IK, Constabel F (eds): "Cell Culture and Somatic Cell Genetics of Plants. Phytochemicals in Cultured Cells," Vol 5. New York: Academic Press, 1988 (in press).

56. Staba EJ, Chung AC: Quinine and quinidine production by *Cinchona* leaf, root and unorganised cultures. Phytochemistry 20:2495, 1981.

57. Anderson LA, Keene AT, Phillipson JD: Alkaloid production by leaf organ, root organ and cell suspension cultures of *Cinchona ledgeriana*. Planta Med 46:25, 1982.

58. Hunter CS, McCalley DV, Barraclough AJ: Alkaloids produced by cultures of *Cinchona ledgeriana* L. In Fujiwara A (ed): "Plant Tissue Culture 1982." Tokyo: Japanese Association for Plant Tissue Culture, 1982, p 317.

59. Koblitz H, Koblitz D, Schmauder H-P, Groger D: Studies on tissue cultures of the genus *Cinchona* L. Alkaloid production in cell suspension cultures. Plant Cell Rep 2:122, 1983.

60. Parr AJ, Robins RJ, Rhodes MJC: Permeabilization of *Cinchona ledgeriana* cells by dimethylsulphoxide. Effects on alkaloid release and long-term membrane integrity. Plant Cell Rep 3:262, 1984.

61. Scragg AH, Morris P, Allan EJ: The effects of plant growth regulators on growth and alkaloid formation in *Cinchona ledgeriana* callus culture. J Plant Physiol 124:371, 1986.

62. Robins RJ, Payne J, Rhodes MJC: Cell suspension cultures of *Cinchona ledgeriana*: 1. Growth and quinoline alkaloid production. Planta Med 52:163, 1986.

63. Payne J, Rhodes MJC, Robins RJ: Quinoline alkaloid production by transformed cultures of *Cinchona ledgeriana*. Planta Med 53:367, 1987.

64. Robins RJ, Hanley AB, Richards SR, Fenwick RG, Rhodes MJC: The production of norharman by suspension cultures of *Cinchona pubescens* fed L-tryptophan. Plant Cell Tissue Organ Culture 9:49, 1987.

65. Schmauder H-P, Groger D, Koblitz H, Koblitz D: Shikimate pathway activity in shake and fermenter cultures of *Cinchona succirubra*. Plant Cell Rep 4:233, 1985.

66. Misawa M, Tanaka H, Mukai N: Production of an anti-tumor substance, camptothecin. Jpn Patent (Kokai) 73:28691, 1973.
67. Sakato K, Misawa M: Effects of chemical and physical conditions on growth of *Camptotheca acuminata* cell cultures. Agric Biol Chem 38:491, 1974.
68. Sakato K, Tanaka H, Mukai N, Misawa M: Isolation and identification of camptothecin from cells of *Camptotheca acuminata* suspension cultures. Agric Biol Chem 38:217, 1974.
69. Rueffer M: The production of isoquinoline alkaloids by plant cell cultures. In Phillipson JD, Roberts MF, Zenk MH (eds): "The Chemistry and Biology of Isoquinoline Alkaloids." Berlin: Springer-Verlag, 1985, p 265.
70. Constabel F: Morphinan alkaloids from plant cell cultures. In Phillipson JD, Roberts MF, Zenk MH (eds): "The Chemistry and Biology of Isoquinoline Alkaloids." Berlin: Springer-Verlag, 1985, p 257.
71. Tanahashi T, Zenk MH: Isoquinoline alkaloids from cell suspension cultures of *Fumaria capreolata*. Plant Cell Rep 4:96, 1985.
72. Akasu M, Itokawa H, Fujita M: Biscoclaurine alkaloids in callus tissues of *Stephania cepharantha*. Phytochemistry 15:471, 1976.
73. Akasu M, Itokawa H, Fujita M: Oxoaporphine alkaloids from callus tissues of *Stephania cepharantha*. Phytochemistry 14:1673, 1975.
74. Ikuta A, Syono K, Furuya T: Alkaloids in plants regenerated from *Coptis* callus cultures. Phytochemistry 14:1209, 1975.
75. Ikuta A, Syono K, Furuya T: Alkaloids of callus tissues and redifferentiated plantlets in the Papaveraceae. Phytochemistry 13:2175, 1974.
76. Furuya T, Yoshikawa T, Kiyohara H: Alkaloid production in cultured cells of *Dioscoreophyllum cumminsii*. Phytochemistry 22:1671, 1983.
77. Ikuta A, Itokawa H: Studies on the alkaloids from tissue culture of *Nandina domestica*. In Fujiwara A (ed): "Plant Tissue Culture 1982." Tokyo: Japanese Association for Plant Tissue Culture, 1982, p 315.
78. Ikuta A, Itokawa H: Berberine and other protoberberine alkaloids in callus tissue of *Thalictrum minus*. Phytochemistry 21:1419, 1982.
79. Gurova TF, Popov YG, Fadeeva II, Shain SS, Tokachev O: Callus formation and accumulation of alkaloids in the tissue culture of *Stephania glabra*. Rastit Resur 16:412, 1980.
80. Khanna P, Sharma OP, Saluja M: Berberine from *Argemone mexicana* Linn. tissue cultures. In Fujiwara A (ed): "Plant Tissue Culture 1982." Tokyo: Japanese Association for Plant Tissue Culture, 1982, p 311.
81. Breuling M, Alfermann AW, Reinhard E: Cultivation of cell cultures of *Berberis wilsoniae* in 20-1 airlift bioreactors. Plant Cell Rep 4:220, 1985.
82. Hinz H, Zenk MH: Production of protoberberine alkaloids by cell suspension cultures of *Berberis* species. Naturwissenschaften 68:620, 1981.
83. Fukui H, Nakagawa K, Tsuda S, Tabata M: Production of isoquinoline alkaloids by cell suspension cultures of *Coptis japonica*. In Fujiwara A (ed): "Plant Tissue Culture 1982." Tokyo: Japanese Association for Plant Tissue Culture, 1982, p 313.
84. Sato F, Yamada Y: High berberine-producing cultures of *Coptis japonica* cells. Phytochemistry 23:281, 1984.
85. Yamamoto H, Nakagawa K, Fukui H, Tabata M: Cytological changes associated with alkaloid production in cultured cells of *Coptis japonica* and *Thalictrum minus*. Plant Cell Rep 5:65, 1986.

86. Lockwood GB: Alkaloids of cell suspensions derived from four *Papaver* spp. and the effect of temperature stress. Z Pflanzenphysiol 144:361, 1984.

87. Iwasa K, Takao N: Formation of alkaloids in *Corydalis ophiocarpa* callus culture. Phytochemistry 21:611, 1982.

88. Koblitz H, Schumann U, Böhm H, Franke J: Tissue cultures of alkaloid plants IV. *Macleaya microcarpa* (Maxin) Fedde. Experentia 31:768, 1975.

89. Staba EJ, Zito S, Amin M: Alkaloid production from *Papaver* tissue cultures. J Nat Prod 45:256, 1982.

90. Anderson LA, Homeyer BC, Phillipson JD, Roberts MF: Dopamine and cryptopine production by cell suspension cultures of *Papaver somniferum*. J Pharm Pharmacol 35:21P, 1983.

91. Kamimura S, Nishikawa M: Growth and alkaloid production of the cultured cells of *Papaver bracteatum*. Agric Biol Chem 40:907, 1976.

92. Furuya T, Ikuta A, Syono K: Alkaloids from callus tissue of *Papaver somniferum*. Phytochemistry 11:3041, 1972.

93. Berlin J, Forche E, Wray V, Hammer J, Hosel W: Formation of benzophenanthridine alkaloids by suspension cultures of *Eschscholtzia californica*. Z Naturforsch 38C:346, 1983.

94. Kutchan TM, Ayabe S, Krueger RJ, Coscia EM, Coscia CJ: Cytodifferentiation and alkaloid accumulation in cultured cells of *Papaver bracteatum*. Plant Cell Rep 2:281, 1983.

95. Khanna P, Sharma GL: Production of opium alkaloids from *in vitro* tissue cultures of *Papaver rhoeas*. Indian J Exp Biol 15:951, 1977.

96. Khanna P, Khanna R: Production of major alkaloids from *in vitro* tissue cultures of *Papaver somniferum* Linn. Indian J Exp Biol 14:628, 1976.

97. Yoshikawa T, Furuya T: Morphinan alkaloid production by opium poppy suspension cultures. In Fujiwara A (ed): "Plant Tissue Culture 1982." Tokyo: Japanese Association for Plant Tissue Culture, 1982, p 307.

98. Tam WHJ, Constabel F, Kurz WGW: Codeine from cell suspension cultures of *Papaver somniferum*. Phytochemistry 19:486, 1980.

99. Hodges CC, Rapoport H: Morphinan alkaloids in callus cultures of *Papaver somniferum*. J Nat Prod 45:481, 1982.

100. Hsu A-F: Effect of protein synthesis inhibitors on cell growth and alkaloid production in cell cultures of *Papaver somniferum*. J Nat Prod 44:408, 1981.

101. Kamo KK, Kimoto W, Hsu A-F, Mahlberg PG, Bills DD: Morphinan alkaloids in cultured tissues and redifferentiated organs of *Papaver somniferum* L. Phytochemistry 21:219, 1982.

102. Hutin M, Foucher JP, Courtois D, Petiard V: Mise en evidence de formes d'accumulation inhabituelles de morphine et de codeine dans une culture de tissus de *Papaver somniferum* L. C R Acad Sci [III] (Paris) 297:47, 1983.

103. Yoshikawa T, Furuya T: Morphinan alkaloid production by tissues differentiated from cultured cells of *Papaver somniferum*. Planta Med 51:110, 1985.

104. Akasu M, Itokawa H, Fujita M: Four new fluorescent components isolated from the callus tissue of *Stephania cepharantha*. Tetrahedron Lett 41:3609, 1974.

105. Kamimura S, Akutsu M, Nishikawa M: Formation of thebaine in the suspension culture of *Papaver bracteatum*. Agric Biol Chem 40:913, 1976.

106. Zito SW, Staba EJ: Thebaine from root cultures of *Papaver bracteatum*. Planta Med 45:53, 1982.

107. Shafiee A, Lalezari I, Yassa N: Thebaine in tissue cultures of *Papaver bracteatum* Lindl. Population Arya II. Lloydia 39:380, 1976.

108. Khanna P, Khanna R, Sharma M: Production of free ascorbic acid and effects of exogenous ascorbic acid and tyrosine on production of major opium alkaloids from *in vitro* tissue culture of *Papaver somniferum*. Indian J Exp Biol 16:110, 1978.

109. Misawa M, Hayashi M, Takayama S: Production of antineoplastic agents by plant tissue cultures. Planta Med 49:115, 1983.

110. Delfel NE, Rothfus JA: Antitumour alkaloids in callus cultures of *Cephalotaxus harringtonia*. Phytochemistry 16:1595, 1977.

111. Delfel NE: The effect of nutritional factors on alkaloid metabolism in *Cephalotaxus harringtonia* tissue cultures. Planta Med 39:168, 1980.

112. Wink M, Witte L, Hartmann T, Theuring C, Volz V: Accumulation of quinolizidine alkaloids in plants and cell suspension cultures: Genera *Lupinus, Cytisus, Baptisia, Genista, Laburnum* and *Sophora*. Planta Med 48:253, 1983.

113. Wink M, Witte L, Schiebel H-M, Hartmann T: Alkaloid pattern of cell suspension cultures and differentiated plants of *Lupinus polyphyllus*. Planta Med 38:238, 1980.

114. Wink M, Schiebel H-M, Witte L, Hartmann T: Quinolizidine alkaloids from plants and their cell suspension cultures. Planta Med 44:15, 1982.

115. Wink M, Hartmann T: Diurnal fluctuation of quinolizidine alkaloid accumulation in legume plants and photomixotrophic cell suspension cultures. Z Naturforsch 37C:369, 1982.

116. Wink M, Hartmann T: The alkaloid patterns of cell suspension cultures and differentiated plants of *Baptisia australis* and their biogenetic implications. J Nat Prod 44:14, 1981.

117. Khanna P, Sharma GL, Uddin A: Atropine from *Atropa belladonna* tissue cultures. Indian J Exp Biol 15:323, 1977.

118. West FR, Mika ES: Synthesis of atropine by isolated roots and root callus cultures of belladonna. Bot Gaz 119:50, 1957.

119. Chan WN, Staba EJ: Alkaloid production by *Datura* callus and suspension tissue cultures. Lloydia 28:55, 1965.

120. Lindsey K, Yeoman MM: The relationship between growth rate, differentiation and alkaloid accumulation in cell cultures. J Exp Bot 34:1055, 1983.

121. Endo T, Yamada Y: Alkaloid production in cultured roots of three species of *Duboisia*. Phtyochemistry 24:1233, 1985.

122. Yamada Y, Endo T: Tropane alkaloid production in cultured cells of *Duboisia leichhardtii*. Plant Cell Rep 3:186, 1984.

123. Kitamura Y, Miura H, Sugii M: Alkaloid composition and atropine esterase activity in callus and differentiated tissues of *Duboisia myoporoides* R. Br. Chem Pharm Bull (Tokyo) 33:5445, 1985.

124. Sipply KJ, Friedrick H: Alkaloide in kallus van *Duboisia myoporoides*. Planta Med Suppl, 1975, p 186.

125. Czygan FC: Moglichkeiten zur produktion von arneistoffen durch pflanzliche gewebe-kulturen. Planta Med Suppl, 1975, p 169.

126. Hashimoto T, Yukimune Y, Yamada Y: Tropane alkaloid production in *Hyoscyamus* root cultures. J Plant Physiol 124:61, 1986.

127. Oksman-Caldentey K-M, Strauss A, Hiltunen R, Vuorela H, Zimmermann W: Variation of scopolamine and hyscyamine content in *Hyoscyamus muticus* L. plants and cell culture clones. In "Proc of 32nd Annu Congr Medicinal Plant Research," Farmaceutisch Tijdschrift voor Belgie, 1984, p 327.

128. Oksman-Caldentey K-M, Strauss A: Somaclonal variation of scopolamine content in protoplast-derived cell culture clones of *Hyoscyamus muticus*. Planta Med 52:6, 1986.

129. Hashimoto T, Sato F, Mino M, Yamada Y: Production of tropane alkaloids from cultured Solanaceae cells. In Fujiwara A (ed): "Plant Tissue Culture 1982." Tokyo: Japanese Association for Plant Tissue Culture, 1982, p 305.

130. Yamada Y, Hashimoto T: Production of tropane alkaloids in cultured cells of *Hyoscyamus niger*. Plant Cell Rep 1:101, 1982.

131. Hashimoto T, Yamada Y: Scopolamine production in suspension cultures and redifferentiated roots of *Hyoscyamus niger*. Planta Med 47:195, 1983.

132. Zheng GZ, Zheng L: Studies on tissue culture of medicinal plants. I Callus cultures of *Scopolia acutangula* for the production of hyoscyamine and scopolamine. Acta Bot Sin 18:163, 1976.

133. Zheng GZ, Zheng L: Studies on tissue cultures of medicinal plants. II: Chemical control of callus growth and synthesis of hyoscyamine and scopolamine by *Scopolia acutangula* callus. Acta Bot Sin 19:209, 1977.

134. Tabata M, Yamamoto H, Hiraoka N, Konoshima M: Organisation and alkaloid production in tissue cultures of *Scopolia parviflora*. Phytochemistry 11:949, 1972.

135. Tiburcio AF, Kaur-Sawhney R, Ingersoll RB, Galston AW: Correlation between polyamines and pyrrolidine alkaloids in developing tobacco callus. Plant Physiol 78:323, 1985.

136. Tiburcio AF, Ingersoll RB, Galston AW: Modified alkaloid pattern in developing tobacco callus. Plant Sci 38:207, 1985.

137. Furuya T, Syono K, Kojima H, Hirotani M, Ikuta A, Hirichi M, Kawaguchi K, Matsumoto K: Chemical constituents and transformation capacity of medicinal plant callus tissues. In "Proc 4th Int Ferm Symp Society of Fermentation Technology," Japan, 1972, p 705.

138. Rhodes MJC, Hilton M, Parr AJ, Hamill JD, Robins RJ: Nicotine production by hairy root cultures of *Nicotiana rustica*: Fermentation and product recovery. Biotechnol Lett 8:415, 1986.

139. Shiio I, Ohta S: Nicotine production by tobacco callus tissues and effect of plant growth regulators. Agric Biol Chem 37:1857, 1973.

140. Speake T, McCloskey P, Smith WK, Scott TA, Hussey H: Isolation of nicotine from cell cultures of *Nicotiana tabacum*. Nature 201:614, 1964.

141. Lockwood GB, Essa AK: The effect of varying hormonal and precursor supplementations on levels of nicotine and related alkaloids in cell cultures of *Nicotiana tabacum*. Plant Cell Rep 3:109, 1984.

142. Hamill JD, Parr AJ, Robins RJ, Rhodes MJC: Secondary product formation by cultures of *Beta vulgaris* and *Nicoiana rustica* transformed with *Agrobacterium rhizogenes*. Plant Cell Rep 5:111, 1986.

143. Keller H, Wanner H, Baumann TW: Kaffein-synthese in fruchtenund gewebekulturen von *Coffea arabica*. Planta 108:339, 1972.

144. Buckland E, Townsley DM: Coffee cell suspension cultures. Caffeine and chlorogenic acid content. J Can Inst Food Sci Technol 8:164, 1975.

145. Frischknecht PM, Baumann TW, Wanner H: Tissue culture of *Coffea arabica* growth and caffeine formation. Planta Med 31:344, 1977.

146. Ogutuga DBA, Northcote DH: Caffeine formation in tea callus tissue. J Exp Bot 21:258, 1970.

147. Ogutuga DBA, Northcote DH: Biosynthesis of caffeine in tea callus tissue. Biochem J 117:715, 1970.

148. Gras M, Creche J, Chenieux JC, Rideau M: Etude comparee des effets de la selection et de facteurs de l'environnement sur l'accumulation alcaloidique de souches de *Choisya ternata*. Planta Med 46:231, 1982.

149. Steck W, Bailey BK, Shyluk JP, Gamborg OL: Coumarins and alkaloids from cell cultures of *Ruta graveolens*. Phytochemistry 10:191, 1971.

150. Steck W, Gamborg OL, Bailey BK: Increased yields of alkaloids through precursor biotransformations in cell suspension cultures of *Ruta graveolens*. Lloydia 36:93, 1973.

151. Boulanger D, Bailey BK, Steck W: Formation of edulinine and furoquinoline alkaloids from quinoline derivatives by cell suspension cultures of *Ruta graveolens*. Phytochemistry 12:2399, 1973.

152. Ramawat KG, Rideau M, Chenieux JC: Growth and quaternary alkaloid production in differentiating and non-differentiating strains of *Ruta graveolens*. Phytochemistry 24:441, 1985.

153. Engel B, Reinhard E: Acridone alkaloid formation in suspension cultures of *Ruta* species. In "Proc of 32nd Annu Congr Medicinal Plant Research," Farmaceutisch Tijdschrift voor Belgie, 1984, p 335.

154. Scharlemann W: Acridin-alkaloide aus kalluskulturen von *Ruta graveolens* L. Z Naturforsch 27C:806, 1972.

155. Baumert A, Kuzovkina IN, Krauss G, Hieke M, Groger D: Biosynthesis of rutacridone in tissue cultures of *Ruta graveolens* L. Plant Cell Rep 1:168, 1982.

156. Nahrstedt A, Wray V, Engel B, Reinhard E: New furoacridone alkaloids from tissue culture of *Ruta graveolens*. Planta Med 51:517, 1986.

157. Roddick JG, Butcher DN: Isolation of tomatine from cultured excised roots and callus tissues of tomato. Phytochemistry 11:2019, 1972.

158. Hertmann E: Biogenesis of steroids. Phytochemistry 22:1843, 1983.

159. Kadkade PG, Madrid TR: Glycoalkaloids in tissue cultures of *Solanum acculeatissimum*. Naturwissenschaften 64:147, 1977.

160. Zacharius RM, Osman SF: Glycoalkaloids in tissue cultures of *Solanum* species. Dehydrocommersonine from cultured roots of *Solanum chacoense*. Plant Sci Lett 10:283, 1977.

161. Emke A, Eilert U: Steroidal alkaloids in tissue cultures and regenerated plants of *Solanum dulcamara*. Plant Cell Rep 5:31, 1986.

162. Willuhn G, May S: Triterpene und steroide in kalluskulturen von *Solanum dulcamara*. Planta Med 46:153, 1982.

163. Jain SC, Khanna P, Sahoo S: *Solanum jasminoides* Paxt., tissue cultures. I. Production of steroidal sapogenins and glycoalkaloids. J Nat Prod 44:125, 1981.

164. Chaturvedi HC, Chowdhury AR, Uddin A: Solasodine biosynthesis in seed- and seedling-callus of *Solanum khasianum* Clarke grown *in vitro*. Indian J Exp Biol 17:107, 1979.

165. Kokate CK, Radwan SS: Enrichment of *Solanum khasianum* callus generating rootlets with steroidal glycoalkaloids. Z Naturforsch 34C:634, 1979.

166. Uddin A, Chaturvedi HC: Solasodine in somatic tissue cultures of *Solanum khasianum*. Planta Med 37:90, 1979.

167. Hosoda N, Yatazawa M: Some accounts on culture conditions of callus tissue on *Solanum laciniatum*. Ait. for producing solasodine. Agric Biol Chem 43:821, 1979.

168. Chandler S, Dodds J: Solasodine production in rapidly proliferating tissue cultures of *Solanum laciniatum* Ait. Plant Cell Rep 2:69, 1983.

169. Chandler S, Dodds J: The effect of phosphate, nitrogen and sucrose on the production of

phenolics and solasodine in callus cultures of *Solanum laciniatum*. Plant Cell Rep 2:205, 1983.

170. Bhatt PN, Bhatt DP: Studies on some factors affecting solasodine contents in tissue cultures of *Solanum nigrum*. Physiol Plant 57:159, 1983.

171. Jain SC, Sahoo S: Isolation and characterisation of steroidal sapogenins and glycoalkaloids from tissue cultures of *Solanum verbascifolium* Linn. Chem Pharm Bull (Tokyo) 29:1765, 1981.

172. Heble MR, Narayanaswamy S, Chadha MS: Solasonine in tissue cultures of *Solanum xanthocarpum*. Naturwissenschaften 55:350, 1968.

173. Khanna P, Uddin A, Sharma GL, Manot SK, Rathore AK: Isolation and characterisation of sapogenin and solasodine from *in vitro* cultures of some solanaceous plants. Indian J Exp Biol 14:694, 1976.

174. Loyola-Vargas VM, Velasco C, Mendez BM, Oropeza C, Reyes J, Robert ML: Biosynthesis of alkaloids in green callus of *Catharanthus roseus*. In Somers DA, Gengenbach BG, Biesboer DD, Hackett WP, Green CE (eds): "VI International Congress of Plant Tissue and Cell Culture." Minneapolis: International Association for Plant Tissue Culture, 1986, p 67.

175. Miura Y, Hirata K, Kurano N: Production of antitumour alkaloids in callus culture of *Catharanthus roseus* (L).G.Don. In Somers DA, Gengenbach BG, Biesboer DD, Hackett WP, Green CE (eds): "VI International Congress of Plant Tissue and Cell Culture." Minneapolis: International Association for Plant Tissue Culture, 1986, p 69.

176. Schubel H, Treiber A, Stöckigt J: Structure revision of the *Rauwolfia* alkaloid raucaffricine. Helv Chim Acta 67:2078, 1984.

177. Khan MA, Ahsan AM: Alkaloids of *Rauwolfia caffra* Sonder. III. Structure of raucaffricine. Pak J Sci Ind Res 15:30, 1972.

178. Habib MS, Court WE: Estimation of the alkaloids of *Rauwolfia caffra*. Planta Med 25:261, 1974.

179. Khan MA, Horn H, Voelter W: Isolation and carbon-13 NMR spectroscopy of indolenine alkaloids. Z Naturforsch 37B:494, 1982.

180. Anderson LA, Phillipson JD, Roberts MF: Biosynthesis of secondary products of cell cultures of higher plants. In Fiechter A (ed): "Advances in Biochemical Engineering/Biotechnology: Plant Cell Culture." Berlin: Springer-Verlag, 1985, Vol 31, p 1.

181. Anderson LA, Hay CA, Roberts MF, Phillipson JD: Studies on *Ailanthus altissima* cell suspension cultures: Precursor feeding of L-[methylene ^{14}C]-tryptophan and L-tryptophan. Plant Cell Rep 5:387, 1986.

182. Anderson LA: Unpublished observations, 1986.

183. Knobloch K-H, Berlin J: Influence of phosphate on the formation of the indole alkaloids and phenolic compounds in cell suspension cultures of *Catharanthus roseus* I. Comparison of enzyme activities and product accumulation. Plant Cell Tissue Organ Culture 2:333, 1983.

184. Merillon JM, Doireau P, Guillot A, Chenieux JC, Rideau M: Indole alkaloid accumulation and tryptophan decarboxylase activity in *Catharanthus roseus* cells cultured in three different media. Plant Cell Rep 5:23, 1986.

185. Zenk MH: Enzymatic synthesis of ajmalicine and related indole alkaloids. J Nat Prod 43:438, 1980.

186. Hemscheidt T, Zenk MH: Partial purification and characterisation of a NADPH dependent tetrahydroalstonine synthase from *Catharanthus roseus* cell suspension cultures. Plant Cell Rep 4:216, 1985.

187. Fahn W, Gundlach H, Deus-Neumann B, Stöckigt J: Late enzymes of vindoline

biosynthesis. Acetyl-CoA: 17-O-deacetylvindoline 17-O-acetyltransferase. Plant Cell Rep 4:333, 1985.

188. Fahn W, LauBermair E, Deus-Neumann B, Stöckigt J: Late enzymes of vindoline biosynthesis. S-adenosyl-L-methionine: 11-O-demethyl-17-O-deacetylvindoline 17-O-methyltransferase and unspecific acetylesterase. Plant Cell Rep 4:337, 1985.

189. Endo T, Goodbody A, Vukovic J, Chapple C, Misawa M, Choi LSL, Kutney JP: Enzymatic synthesis of 3′, 4′ − anhydrovinblastine by cell free extracts from cultured *Catharanthus roseus* cells. In Somers DA, Gengenbach BG, Biesboer DD, Hackett WP, Green CE (eds): ''VI International Congress of Plant Tissue and Cell Culture.'' Minneapolis: International Association for Plant Tissue Culture, 1986, p 143.

190. Uesata S, Kanomi S, Iida A, Inouye H, Zenk MH: Mechanism for iridane skeleton formation in the biosynthesis of secologanin and indole alkaloids in *Lonicera tatarica*, *Catharanthus roseus* and suspension cultures of *Rauwolfia serpentina*. Phytochemistry 25:839, 1986.

191. Pfitzner A, Krausch B, Stöckigt J: Characteristics of vellosimine reductase, a specific enzyme involved in the biosynthesis of the *Rauwolfia* alkaloid sarpagine. Tetrahedron 40:1691, 1984.

192. Pfitzner A, Polz L, Stöckigt J: Properties of vinorine synthase—the *Rauwolfia* enzyme involved in the formation of the ajmaline skeleton. Z Naturforsch 41C:103, 1986.

193. Schubel H, Stöckigt J, Feicht R, Simon H: Partial purification and characterisation of raucaffricine-β-D-glucosidase from plant cell suspension cultures of *Rauwolfia serpentina* Benth. Helv Chim Acta 69:538, 1986.

194. Hay CA, Anderson LA, Roberts MF, Phillipson JD: *In vitro* cultures of *Cinchona* species. Precursor feeding of *C. ledgeriana* root organ suspension cultures with L-tryptophan. Plant Cell Rep 5:1, 1986.

195. Hay CA: Unpublished observations, 1986.

196. Skinner SE, Walton NJ, Robins RJ, Rhodes MJC: Tryptophan decarboxylase, strictosidine synthase and alkaloid production by *Cinchona ledgeriana* Moens suspension cultures. Phytochemistry 26:721, 1987.

197. Isaac JE, Robins RJ, Rhodes MJC: Cinchoninone: NADPH oxidoreductases I and II— novel enzymes in the biosynthesis of quinoline alkaloids in *Cinchona ledgeriana*. Phytochemistry 26:393, 1987.

198. Zenk MH, Rueffer M, Amann M, Deus-Neumann B: Benzylisoquinoline biosynthesis by cultivated plant cells and isolated enzymes. J Nat Prod 48:725, 1985.

199. Wat C-K, Steffens P, Zenk MH: Partial purification and characterisation of S-adenosyl-L-methionine: Norrecticuline N-methyltransferases from *Berberis* cell suspension cultures. Z Naturforsch 41C:126, 1986.

200. Steffens P, Nagakura N, Zenk MH: Purification and characterisation of the berberine bridge enzyme from *Berberis beaniana* cell cultures. Phytochemistry 24:2577, 1985.

201. Muemmler S, Rueffer M, Nagakura N, Zenk MH: S-adenosyl-L-methionine: (S)-scoulerine 9-O-methyltransferase, a highly stereo- and regio-specific enzyme in tetrahydroprotoberberine biosynthesis. Plant Cell Rep 4:36, 1985.

202. Rueffer M, Amann M, Zenk MH: S-adenosyl-L-methionine: Columbamine-O-methyltransferase, compartmentalized enzyme in protoberberine biosynthesis. Plant Cell Rep 3:182, 1986.

203. Rueffer M, Zenk MH: Columbamine, the central intermediate in the late stages of protoberberine biosynthesis. Tetrahedron Lett 27:923, 1986.

204. Yamada Y, Okada N: Biotransformation of tetrahydroberberine into berberine by enzymes prepared from cultured *Coptis japonica* cells. Phytochemistry 24:63, 1985.

205. Amann M, Wanner G, Zenk MH: Intracellular compartmentation of two enzymes of berberine biosynthesis in plant cell cultures. Planta 167:310, 1986.
206. Wink M: Metabolism of quinolizidine alkaloids in plants and cell suspension cultures: Induction and degradation. In Neumann D (ed): "Primary and Secondary Metabolism of Plant Cell Cultures." Berlin: Springer-Verlag, 1985, p 107.
207. Hashimoto T, Yamada Y: Hyoscyamine 6-β-hydroxylase, a 2-oxoglutarate-dependent dioxygenase, in alkaloid producing root cultures. Plant Physiol 81:619, 1986.
208. Baumert A, Kuzovkina IN, Gröger D: Activation of anthranilic acid and N-methyl-anthranilic acid by cell-free extracts from *Ruta graveolens* tissue cultures. Planta Med 51:125, 1985.
209. Baumert A, Schneider G, Gröger D: Biosynthesis of acridone alkaloids. A cell-free system from *Ruta graveolens* cell suspension cultures. Z Naturforsch 41C:187, 1986.
210. Fairbairn JW, Hakim F, El Kheir Y: Alkaloidal storage metabolism and translocation in the vesicles of *Papaver somniferum* latex. Phytochemistry 13:1133, 1974.
211. Wink M: Storage of quinolizidine alkaloids in epidermal tissues. Z Naturforsch 41C:375, 1986.
212. Eilert U, Wolters B, Constabel F: Ultrastructure of acridone alkaloid idioblasts in roots and cell cultures of *Ruta graveolens*. Can J Bot 64:1089, 1986.
213. Mersey BG, Cutler AJ: Differential distribution of specific indole alkaloids in leaves of *Catharanthus roseus*. Can J Bot 64:1039, 1986.
214. Lang H, Kohlenbach HW: Differentiation of alkaloid cells in cultures of *Macleaya* mesophyll protoplasts. Planta Med 46:78, 1982.
215. Von Borstel K, Hartmann T: Selective uptake of pyrrolizidine N-oxides by cell suspension cultures from pyrrolizidine alkaloid producing plants. Plant Cell Rep 5:39, 1986.
216. Roberts MF: *Papaver*. In Vasil IK, Constabel F (eds): "Cell Culture and Somatic Cell Genetics of Plants." New York: Academic Press, Vol 5, 1988, p 315.
217. Nessler CL: Somatic embryogenesis in the opium poppy, *Papaver somniferum*. Physiol Plant 55:453, 1982.
218. Schuchmann R, Wellmann E: Somatic embryogenesis in tissue cultures of *Papaver somniferum* and *Papaver orientale* and its relationship to alkaloid and lipid metabolism. Plant Cell Rep 2:88, 1983.
219. Kutchan TM, Ayabe S, Coscia CJ: Cytodifferentiation and *Papaver* alkaloid accumulation. In Phillipson JD, Roberts MF, Zenk MH (eds): "The Chemistry and Biology of Isoquinoline Alkaloids." Berlin: Springer-Verlag, 1985, p 281.
220. Neumann D, Krauss G, Hieke M, Gröger D: Indole alkaloid formation and storage in cell suspension cultures of *Catharanthus roseus*. Planta Med 48:20, 1983.
221. Brown S, Renaudin JP, Prevot C, Guern J: Flow cytometry and sorting of plant protoplasts: Technical problems and physiological results from a study of pH and alkaloids in *Catharanthus roseus*. Physiol Veg 22:541, 1984.
222. Brown S: Analysis and sorting of plant material by flow cytometry. Physiol Veg 22:341, 1984.
223. Matile P: Das toxische Kompartiment der Pflanzenzelle. Naturwissenschaften 71:18, 1984.
224. Homeyer BC, Roberts MF: Alkaloid sequestration by *Papaver somniferum* latex. Z Naturforsch 39C:876, 1984.
225. Renaudin JP, Guern J: Compartmentation mechanisms in indole alkaloids in cell suspension cultures of *Catharanthus roseus*. Physiol Veg 20:533, 1982.
226. Renaudin JP, Guern J: Compartmentation of alkaloids in a cell suspension of *Catharan-*

thus roseus: A reappraisal of the role of pH gradients. In Neumann D (ed): "Proceedings of the Symposium on Primary and Secondary Metabolism of Plant Cell Cultures." Berlin: Springer-Verlag, 1985, p 192.

227. Deus-Neumann B, Zenk MH: A highly selective alkaloid uptake system in vacuoles of higher plants. Planta 162:250, 1984.

228. Matile P: Biochemistry and function of vacuoles. Annu Rev Plant Physiol 29:193, 1978.

229. Kurkdjian A: Absorption and accumulation of nicotine in *Acer pseudoplatanus* and *Nicotiana tabacum* cells. Physiol Veg 20:73, 1982.

230. Renaudin JP: Uptake and accumulation of an indole alkaloid [^{14}C]-tabernanthine, by cell suspension cultures of *Catharanthus roseus* (L) G. Don., and *Acer pseudoplatanus* L. Plant Sci Lett 22:59, 1981.

231. Deus-Neumann B, Zenk MH: Accumulation of alkaloids in plant vacuoles does not involve an ion-trap mechanism. Planta 167:44, 1986.

232. Roberts MF, Homeyer BC: Unpublished observations.

233. Wink M: Evidence for an extracellular lytic compartment in plant cell suspension cultures: The cell culture medium. Naturwissenschaften 71:635, 1984.

234. Merillon JM, Chenieux JC, Rideau M: Cinetique de croissance evolution du metabolisme glucido-azote et accumulation alkaloidique dans un suspension cellulaire de *Catharanthus roseus*. Physiol Veg 47:169, 1983.

235. Petiard V: Mise en evidence d'alcaloides dans le milieu nutritif de cultures de tissues de *Catharanthus roseus* (L.) G. Don. Physiol Veg 18:331, 1980.

236. Vinas R, Pareilleux A: Production d'alcaloides par des supensions cellulaires de *Catharanthus roseus* cultivees *in vitro*. Physiol Veg 20:219, 1982.

237. Böhm H: Regulation of alkaloid production in plant cell cultures. In Thorpe TA (ed): "Frontiers of Plant Tissue Culture 1978." Calgary: International Association for Plant Tissue Culture, 1978, p 201.

238. Frischknecht PM, Baumann TW: The pattern of purine alkaloid formation in suspension cultures of *Coffea arabica*. Planta Med 40:245, 1986.

239. Kibler R, Neumann KH: Alkaloidgehalte in haploiden und diploiden blättern und Zellsuspension von *Datura innoxia*. Planta Med 35:354, 1979.

240. Tsukada M, Tabata M: Intracellular localization and secretion of naphthoquinone pigments in cell cultures of *Lithospermum erythrorhizon*. Planta Med 50:338, 1984.

241. Fujita Y, Hara Y: The effective production of shikonin by cultures with an increased cell population. Agric Biol Chem 49:2071, 1985.

242. Fowler MW: Plant cell culture and natural product synthesis: An academic dream or a commercial possibility? Bioessays 3:172, 1985.

243. Curtin ME: Harvesting profitable products from plant tissue culture. Biotechnology 1:649, 1983.

244. Pfitzner U, Zenk MH: Immobilization of strictosidine synthase from *Catharanthus* cell cultures and preparative synthesis of strictosidine. Planta Med 46:10, 1982.

245. Mantell SH, Smith H: Cultural factors that influence secondary metabolite accumulations in plant cell and tissue cultures. In Mantell SH, Smith H (eds): "Plant Biotechnology." Cambridge: Cambridge University Press, 1983, p 75.

246. DiCosmo F, Towers GHN: Stress and secondary metabolism in cultured plant cells. In Timmermann BN, Steelink C, Loewus FA (eds): "Phytochemical Adaptations to Stress." New York: Plenum Press, 1984, p 97.

247. Misawa M: Production of useful plant metabolites. In Fiechter A (ed): "Advances in Biochemical Engineering/Biotechnology: Plant Cell Culture." Berlin: Springer-Verlag, 1985, Vol 31, p 59.

248. Siebert M, Kadkade PG: Environmental factors. In Staba EJ (ed): "Plant Tissue Culture as a Source of Biochemicals." Boca Raton: CRC Press Inc, 1980, p 123.

249. Martin SM: Environmental factors. In Staba EJ (ed): "Plant Tissue Culture as a Source of Biochemicals." Boca Raton: CRC Press Inc, 1980, p 143.

250. DiCosmo F, Misawa M: Eliciting secondary metabolism in plant cell cultures. Trends Biotechnol 3:318, 1985.

251. Eilert U, Ehmke A, Wolters B: Elicitor-induced accumulation of acridone alkaloid epoxides in *Ruta graveolens* suspension cultures. Planta Med 50:508, 1984.

252. Rhodes MJC: Immobilized plant cell cultures. Top Enzyme Ferment Biotechnol 10:51, 1985.

253. Yeoman MM: Immobilized plant cells. In Yeoman MM (ed): "Plant Cell Culture Technology." Oxford: Blackwell, 1986, p 229.

254. Brodelius P: Immobilized plant cells and protoplasts. In Evans DA, Sharp WR, Amminato PV, Yamada Y (eds): "Handbook of Plant Cell Culture." New York: Macmillan, 1986, p 287.

255. Deus-Neumann B, Zenk MH: Instability of indole alkaloid production in *Catharanthus roseus* cell suspension cultures. Planta Med 50:427, 1984.

256. Robins RJ, Webb AJ, Rhodes MJC, Payne J, Morgan MRA: Radioimmunoassay for the quantitative determination of quinine in cultured plant tissues. Planta Med 50:235, 1984.

257. Deus B, Zenk MH: Exploitation of plant cells for the production of natural compounds. Biotechnol Bioeng 24:1965, 1982.

258. Fecker LF, Beier H, Berlin J: Cloning and characterization of a lysine decarboxylase gene from *Hafnia alvei*. Mol Gen Genet 203:177, 1986.

Biotechnology in Agriculture, pages 141–174
© 1988 Alan R. Liss, Inc.

Potential for Exploiting Vesicular-Arbuscular Mycorrhizas in Agriculture

I.R. Hall

Invermay Agricultural Centre, Ministry of Agriculture and Fisheries, Private Bag, Mosgiel, New Zealand

———◆◆———

———◆◆———

Vesicular-arbuscular mycorrhizas (VAM) can improve plant growth by stimulating the uptake of nutrients, especially phosphorus (Fig. 1) [43, 97,168], suppressing the detrimental effects of root pathogens [18,35,36,78] (Table I) and possibly by having beneficial effects on plant hydration (see Table XV:ii for references). However, the beneficial effects isolates and species of VAM fungi have on plant growth can vary [4,5,6,50,58,61,

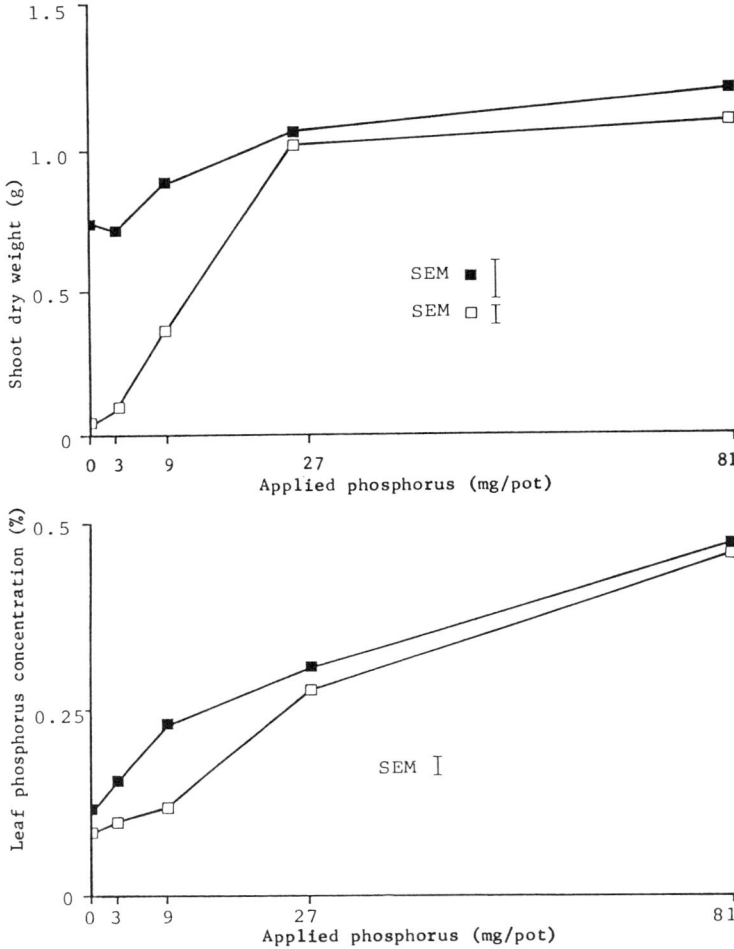

Fig. 1. Effect of a mixed inoculum of *Glomus fasciculatum* and *Gigaspora margarita* on Grasslands Huia white clover (*Trifolium repens* L.) growth in a phosphorus-deficient steamed soil in a greenhouse pot experiment. □, uninoculated; ■, inoculated [from 86].

62,74,118,125,131,133,163,166,185,203,223]. This plus the demonstration that some soils contain few or no VAM fungi, or are populated by VAM fungi less effective than those which can be introduced (Table II; see sections I and II.A for references) has attracted the attention of applied research workers. This paper reviews this applied research and discusses the possi-

TABLE I. Effect of a Mixed VAM Inoculum, *Meloidogyne hapla* Chitwood and Applied Phosphate on Shoot Dry Weight of Wairau Alfalfa (*Medicago sativa* L.) and Number of Nematodes Per Gram of Root [From 78—VAM and Nematodes Added at Transplanting]

Inoculum		Added phosphorus (kg P/ha)			
VAM	Nematodes	0	8	30	120
Shoot dry weight (g—square-root transformed)					
−	−	4.24	4.69	9.27	13.71
+	−	10.95	10.77	12 .77	17.44
−	+	3.16	3.46	5.83	11.49
+	+	11.31	9.70	13.82	16.73
			LSD (5%) 1.08		
Nematodes/g fresh weight of root (square-root transformed)					
−	+	44 .72	37.42	46.90	44.72
+	+	14.14	14.14	28. 28	24.49
			LSD (5%) <9.14		

TABLE II. Effect of Two VAM Inocula on Soybean [*Glycine max* (L.) Merrill cv. Tainon 4] Seed Yield (g/pot) Grown for 84 Days in Four Lowland Subtropical Soils in the Greenhouse [From 230: Experiment 1]

	Soil			
Inoculum	Taichung	Pingtung	Changhua	Tainan
Glomus fragile	4.7	8.9	6.5	4.9
G. fasciculatum	4.2	10. 0	6.9	5.3
None	3.9	7.5	4.8	3.7
		LSD (5%) 0.8		

bilities and practicalities of exploiting VAM symbiosis in agriculture. It has therefore been written more from an agriculturalist's point of view than a mycologist's, and readers who would prefer a different emphasis are referred to Harley and Smith [97], papers in the book edited by Powell and Bagyaraj [184], and papers by Mosse [157] and Smith and Gianinazzi-Pearson [211].*

*This chapter was prepared before the 7th North American Conference on Mycorrhizas. There are a number of papers presented in the Proceedings which are relevant to this review.

I. POT EXPERIMENTS

Pot experiments in sterilized soil have proved valuable by providing much useful information, for example, on the role VAM play in the mineral nutrition (Fig. 1) [8,43] and carbon economy [98,220] of plants, and by demonstrating apparent differences in the effectiveness of VAM fungi (Table II; see above for references). Partial or complete soil sterilization, however, can change its nutrient status and structure [116,142,160,195], as well as removing some or all of the microbiota [30,160], while VAM fungi appear to be adapted to such specific soil conditions. The results of experiments in sterilized soils which purport to show that a soil's indigenous VAM fungi differed in effectiveness from others with which it was compared [e.g., 175,180] are therefore questionable because the indigenous fungi may no longer be well adapted to the changed soil conditions. The same criticisms do not apply to similar greenhouse pot experiments which used unsterile soils containing their normal complement of VAM fungi [2,38,91,135,156, 176,230]. However, these experiments too are open to question. For example, Young and co-workers' [230] (Table II) data could be interpreted as showing that both *Glomus fragile* (Berk. & Broome) Trappe & Gerd. and *G. fasciculatum* (Thaxter sensu Gerd.) Gerd. & Trappe were superior VAM fungi to the indigenous ones in the four soils. It is conceivable, however, that the inoculum potential of the four soils may have been naturally low or depressed by prolonged storage (see section II.A.1). This could have resulted in the delayed onset and benefits of infection in the controls, which could have been particularly marked in relatively short-term experiments.

Another limitation of pot experiments is that greenhouse environments usually differ from those in the field. For example, water is not usually allowed to become limiting in greenhouse studies [cf. 32], but tolerance to drought may be the most important factor to which a VAM fungus may have to be adapted [3]. Plants raised in containers in the greenhouse and then placed in or on the soil surface in the field offer some compromise, but where the effects of VAM in agriculture are under study there is no substitution for field experiments conducted using standard agricultural practices and with soil and climatic conditions, etc., as close as possible to those to which a crop/pasture is normally exposed.

II. FIELD EXPERIMENTS

Despite all the work that has been done on VAM over the past two decades regrettably only a limited amount has been directly aimed at exploiting the symbiosis in agriculture and relatively few researchers have attempted field experiments. There are a number of reasons for this omission:

- There are only a limited number of establishments where VAM research is being conducted which have the facilities and expertise for conducting field experiments.
- Ph.D. and M.Sc. supervisors are unwilling to let their students conduct risky field investigations and have therefore steered them towards greenhouse and laboratory work.
- Many institutions involved in VAM research would find the cost of extensive field experiments prohibitive. For example, I estimate that the current cost of my own series of field experiments and preliminary experiments [89,90] would be in excess of $200,000.
- Much VAM research is conducted simply to further our scientific understanding of the symbiosis, and commercial values have little relevance.

A. Field-Sown Crops and Pastures

In some pot and field experiments dealing with normally field-sown crop and pasture species, a comparison of the effectiveness of indigenous and introduced VAM fungi was made by transplanting from the glasshouse to the field uninoculated seedlings (controls) or seedlings inoculated with an ''elite'' strain of a VAM fungus and then comparing subsequent plant growth [17,101,105,113,115,123,124,197]. Such experiments have two inherent errors:

1. Plants inoculated before transplanting would benefit from mycorrhizas for the period from inoculation to transplanting, while the uninoculated controls would not. For example, in Khan's [123,124] experiments this time advantage was from 18% to 20% of the total experimental period and in Islam and colleagues' [112] from 20% to 43%. At transplanting this advantage may not have been apparent, but by the end of the experiment it would have at least contributed to significant differences in treatments.

2. Work by Hall [83] suggests that as uninoculated seedlings get older they respond to VAM infection more slowly, and hence the controls would have been further disadvantaged.

In other experiments using pre-inoculated transplant techniques these criticisms were overcome by inoculating the controls with a culture of the indigenous fungi from the soil into which the seedlings were later to be transplanted [e.g., 176,178,179]. However, even in these experiments

TABLE III. Effect on Plant Height (Square-Root Transformed) of Inoculating Pelletted Field Grown *Lotus pedunculatus* **Cav. cv. Grasslands Maku With** *Glomus fasciculatum* **[From 88]**

Inoculum	Applied P (kg/ha)	
	10	50
Non-VAM-infested pellet	2.28	2.81
VAM-infested pellet	2.95	3.52
	LSD (5%) 0.34	

competition among the indigenous VAM fungi, the soil flora and fauna, and the inoculant fungus develops only after transplanting. All the phases of growth and competition that an inoculant VAM fungus might otherwise have encountered from the initiation of hyphal growth from a resting propagule, growth through the soil to a root, and the formation of pre-infecting structures followed by infection are bypassed [3]. Also, transplanting infected seedlings to inoculate field-sown crops and pastures with VAM fungi is not a practical technique (see section IV), and consequently the results of experiments which employed these techniques for normally field-sown crops and pastures must be interpreted with some caution.

Two field techniques which have been used to investigate the effects of VAM on the growth of crops in the field are comparing the growth of plants in unfumigated soil with that in fumigated soil [171,222,229], and removing the indigenous fungi with soil fumigation followed by re-inoculation of half of the plots [33,34,115,117,192,201]. However, in the former the beneficial (removal of pathogens and release of nutrients) and detrimental (e.g., bromine residues) effects of fumigation can be confounded, with the loss of the potential benefits of VAM fungi [also see 229]. In addition, neither type of experiment can determine whether inoculant fungi were any more effective than the indigenous ones under normal soil conditions. To study this it is necessary to conduct field experiments in which normal agronomic practices have been followed and, preferably, inocula applied using techniques which could be adapted to agriculture.

1. Soils with low VAM inoculum potentials. Subsoils, eroded soils, fumigated soils, and mine spoils can contain low VAM fungal populations [10,71,82,93,116,153,189,191,225]. It also seems likely that those VAM fungi which are present in these soils would have been derived from miscellaneous accidental natural introductions and therefore may not be those best suited to the conditions. The beneficial effects of VAM inocula on plant growth demonstrated in Hall's [89] (Table III) experiment on an eroded soil, Haas and co-workers' [82] on a methyl bromide fumigated soil,

TABLE IV. Effect of *Glomus macrocarpum* and Fertilizer Phosphorus on Yield and VAM Infection of Potato (*Solanum tuberosum* L.) Growing in the Field in a Fallowed Soil [From 28]

	Triple superphosphate application (kg/ha)	Tuber yield (t/ha)	VAM infection in July (%)
Inoculated with *G. macrocarpum*	0	8.07	24.3
	481	9.62	2.6
Not inoculated	0	6.72	4.6
	481	10.37	1.6
		<1.54	<19.7
		LSD (5%)	

TABLE V. Effect of VAM Inocula on Mean Shoot Dry Weights (g/m Row; Log_e Transformed) of Lucerne cv. European, Onion (*Allium cepa* L. cv. Ailsa Craig) and Barley (*Hordeum vulgare* L. cv. Ark Royal) Grown in Unsterile Field Soil [From 167]

Inoculum	Crop		
	Lucerne	Onion	Barley
Glomus mosseae + other VAM fungi	0.858	−0.139	2.178
Glomus c. caledonium	2.145	0.972	2.213
None	0.031	−1.663	1.665
		LSD (5%) 0.748	

Swaminathan and Verma's [216] on Phagu reclaimed terrace soil, and Hashim-Chulan [99] and Lambert and Cole's [138] on mine spoils were therefore not unexpected. In Hall's [88] experiment the soil was also very deficient in phosphorus and had a very high phosphorus sorbing capacity; and the host plant, *Lotus pedunculatus* Cav. cv. Grasslands Maku, was one known to be reliant on the formation of VAM for vigorous growth in phosphorus-deficient soils [93,181]. Similarly, the experiments of Haas et al. [82] on *Capsicum* (bell pepper) were conducted in a very high phosphorus sorbing soil which had been fumigated to control pepper collapse disease and therefore contained a low VAM fungal density.

The fallowed soil used by Black and Tinker [28] in their study on potatoes was also clearly deficient in VAM fungi, as evidenced by the low infection levels in the uninoculated plots (Table IV). This accounted for the observed responses to inoculation in the absence of applied phosphorus. Similarly in Owusu-Bennoah and Mosse's [167] experiment (Table V), even though by the end of the experiment infection levels were high in the uninoculated controls, the responses to inoculation were surprisingly large, suggesting that the previously fallowed soil either was deficient in VAM fungi or was populated by relatively ineffective species. Also, the response to inoculation

TABLE VI. Effect of *Glomus fasciculatum* **Inoculum on Soybean (cv. Hardee) Growing in a Fallowed Soil [From 19]**

Inocula	Shoot dry wt/ plant(g)	Grain yield/ 1.2 m² plot(g)	Shoot N content (mg)	Shoot P content (mg)	Nodule dry wt/ plant(g)	45 days after inoculation	
						VAM infection level(%)	VAM spores/ 50ml soil
Rhizobium alone	2.83	85.0	93.4	11.3	0.24	65	226
Rhizobium + *G.fasciculatum*	4.65	100.7	198.1	26.6	0.41	82	274
LSD(5%)	<1.82	>15.73	<104.7	<15.38	<0.17	na	<48

na = not analyzed

TABLE VII. Response of Field Grown Subterranean Clover (*Trifolium subterraneum* L. cv. Seaton Park) to Inoculation With *Glomus fasciculatum* **[From 9—Site 2, Harvest 1]**

Inoculum	Individual plant weight (g)	Total infection (%)
None	0.24	10
G.fasciculatum	0.42	42
LSD(5%)	<0.08	<10

(Table VI) produced in soybean [19] and in barley [179] in fallowed soils could have been due to a low VAM fungal density.

2. Pastures. The primary aim in the experiments of Abbott et al. [9] was not to determine if growth could be stimulated by introducing more effective strains of VAM fungi but to investigate the possibility of establishing inoculant fungi in soils which were already well populated by indigenous VAM fungi. Two inoculant fungi, *Glomus fasciculatum* and *G. monosporum* Gerd. & Trappe, were compared at four experimental sites chosen on the basis of the infectivity of their indigenous VAM fungi. The soils ranged from severely phosphorus deficient to well fertilized and not responsive to phosphorus. The *G. monosporum* inoculum failed, but *G. fasciculatum* did establish and raise infection levels at two sites. At the less fertile of these two sites this increase in infection level was accompanied by a transitory growth response at early harvests (Table VII).

Other experiments conducted by Azcon-Aguilar and Barea [16], Hall [90], Hayman [102,103], Newbould and Rangeley [162], and Rangeley et al. [188] were designed specifically to determine if pasture growth could be

TABLE VIII. Effect of VAM Inoculation on the Yield of Grasslands Huia White Clover (kg Dry Matter/ha) in an Acidic Brown Earth [From 188]

	Applied P(kg/ha)	
Inoculum	0	40
None	1,070	841
Glomus etunicatum	598	1,930
	LSD (5%) 572	

TABLE IX. Effect of VAM Inocula on Pasture Dry Matter Yields (t/ha) on a Soil With 15 μg Olsen Available P/ml Meaned Over Four Levels of Applied P [From 90: Experiment 3]

	Year 1			
Inoculum	Harvest 1	Harvest 2 + 3	Year 2	Year 3
Glomus mosseae	0.80	1.12	6.32	6.15
Glomus macrocarpum	0.91	1.16	6.08	6.01
Glomus tenue + *G. pallidum*	0.93	1.14	5.78	5.92
None	0.92	1.02	5.56	5.59
5% LSD	0.10	0.07	0.35	0.40

stimulated by inoculant VAM fungi. The soils on which their experiments were conducted ranged from very infertile hill country soils to high-fertility alluvial flats. In some of the experiments, rate of applied phosphorus was a treatment factor and some were carried on for more than 1 yr. The most widely used inoculant fungus was *Glomus mosseae* (Nicol. & Gerd.) Gerd. & Trappe, but a number of others were also employed. Proportionately the largest responses to inoculation were detected in the less fertile soils (Tables VIII, IX). For example, in Rangeley and co-workers' [188] (Table VIII) experiment on an acidic brown earth in Roxburghshire, United Kingdom, clover yields were doubled (ca. 1,000 kg dry matter/ha), providing 40 kg/ha phosphorus was also applied, while in Hall's experiment on a high-fertility alluvial soil [90: experiment 4], the maximum increase was only 5% (640 kg dry matter/ha). On this highly fertile soil, responses to inoculation occurred in the only year there was also a response to fertilizer phosphorus. Similarly, in an experiment on another fertile site which was accidentally top-dressed with about 50 kg/ha phosphorus on two occasions, no response to phosphorus or inoculation could be detected (Hall, unpublished data). A reduced response to VAM inocula with increasing soil phosphorus levels is in keeping with the results of VAM/phosphorus response curve pot experiments (Fig. 1) [1,29,86,181]. Even so, the tissue phosphorus levels at which responses

TABLE X. Effect of Applied Phosphorus and *Glomus mosseae* Inoculation on Growth and Seed Yield of Broad Bean (*Vicia faba* L.) Grown in Buried Cylinders in the Field [From 136 With Site Used as the Replication Factor]

	Applied phosphorus (μg P/g soil)		
Inoculum	0	Medium (5.5 to 9)	High (11 to 18)
Shoot dry weight (g/cylinder at 8 weeks)			
Glomus mosseae	17.3	21.4	22.5
None	13.6	20.4	18.8
		LSD(5%) (inoculum) 2.67	
		LSD(5%) (phosphorus) 3.27	
Seed dry weight (g/cylinder at harvest)			
Glomus mosseae	8.3	14.0	16.3
None	6.2	10.8	12.4
		LSD(5%) (inoculum) 2.64	
		LSD(5%) (phosphous) 3.23	

occurred on Hall's [90: experiment 4] most fertile site were very high (0.4%), indicating that the beneficial effects of the inoculant VAM fungi may not have been restricted to merely utilizing fertilizer phosphorus more efficiently.

3. Field-sown crops. Field inoculation experiments on field-sown crops on apparently normal agricultural soils have been conducted on soybean [*Glycine max* (L.) Merr. [63,137], cereals [33,34,39,187], faba beans [136], onions [183], and cotton [190]. In these experiments, maximum responses to inoculation were up to 35% (Table X) [136], but in none of these experiments had there been any preselection of fungi likely to produce the maximum growth responses under the conditions of the experiments. Had this been done, it is conceivable that the responses to inoculation would have been greater. In those experiments where rates of phosphorus had been applied, proportional responses to inoculation generally decreased with increasing level of applied phosphorus (Fig. 2).

B. Transplanted Crops—Seedlings Raised in "Sterile" Media

Many horticultural crops and ornamental species are raised in fumigated or heat-treated soil or in essentially sterile soilless potting media. The principal reasons for this are that losses from pests and pathogens are reduced and plant growth rates can be more predictable. Also, seedlings of some species can be raised aseptically using tissue culture techniques [228]. But the growth of plants in these media can be very poor owing to the detrimental effects heat and fumigants can have on VAM or the absence of VAM from

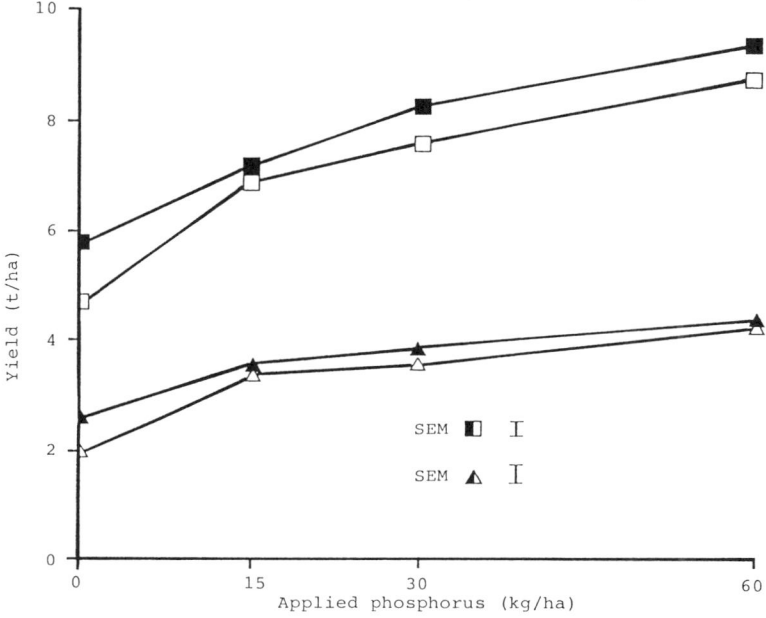

Fig. 2. Effect of *Glomus mosseae* inoculum applied below the seed on spring wheat (*Triticum aestivum* L.) grain (triangles) and total (squares) yields. □ △, uninoculated; ■ ▲, inoculated [from 33].

soil-less potting mixes [144,148,170]. The problem can therefore be rectified either by inoculating with VAM fungi or relieving the limiting factor that VAM would normally help correct, for example, by the application of phosphorus. Some species in which poor growth in media devoid of VAM fungi has been remedied either by inoculating with VAM fungi or by applying nutrients are—

- Apple [65,108,172]
- Avocado [146]
- Bell pepper [130]
- Cassava [109]
- Citrus [77,147]
- Grapes [151]
- Onions [212]
- Peach [141]
- Raspberry [154]

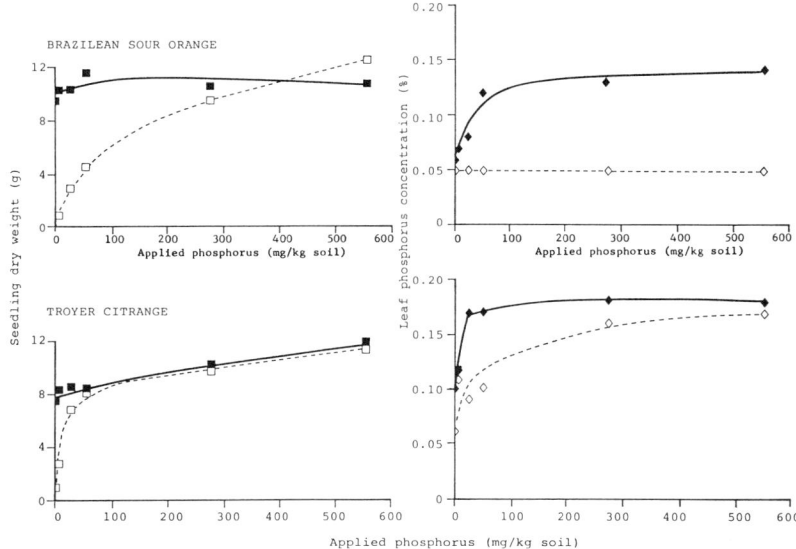

Fig. 3. Effect of *Glomus fasciculatum* inoculum and fertilizer phosphorus on seedling dry weight (squares) and percent phosphorus (diamonds) in leaves of Brazilean sour orange (*Citrus aurantium* L.) and Troyer citrange [*Poncirus trifoliata* (L.) Raf. X *Citrus sinensis* (L.) Osbeck] growing in phosphorus-deficient (4.6 µg/g bicarbonate extractable) autoclaved soil. □ ◇, uninoculated; ■ ◆, inoculated [from 149].

- Tamarillo [42]
- Forest species and ornamentals [13,23,27,45,46,120,127,129,143, 182,215]

For some plant and soil combinations and where VAM normally stimulates phosphorus uptake, the amount of phosphorus which has to be applied to nonmycorrhizal plants in order to get them to grow as well as mycorrhizal ones can be very large. For example, when Brazilian sour orange was grown in a low-fertility soil (4.6 µg available phosphorus/ml soil) approximately 8 t/ha of single superphosphate had to be applied to uninoculated plants to stimulate growth and tissue phosphorus concentrations to the level of unfertilized inoculated ones (Fig. 3) [149]. Similarly, Howeler et al. [109], working on cassava, found that nonmycorrhizal plants had to be fertilized with 1,600 kg phosphorus/ha for their growth rate to reach that of mycorrhizal plants receiving no additional phosphorus. But it should be realized that these are extreme examples resulting from the use of hosts highly dependent on VAM and soils very deficient in available phosphorus.

TABLE XI. Effect of *Glomus mosseae* **and Applied Phosphorus on the Shoot Dry Weight of Apple (***Malus domestica* **Borkh.) Seedlings Grown in Fumigated Soils [From 108]**

			Soil			
	VAM inoculum	Okanogan loam	Taunton fine silty loam	Magallan fine sandy loam	Chelan gravelly sandy loam	Esquatzel silt loam
Extractable phosphorus (μg/g soil)		1.6	11.5	12.2	19.4	60.6
pH		6.0	7.8	6.2	6.4	7.3
Applied phosphorus (mg P/ kg soil)[a]						
0	−	0.6	0.4	0.7	0.5	3.9
	+	1.6	2.7	1.0	1.5	3.9
50	−	3.2	1.4	2.1	1.9	3.8
	+	3.5	3.6	2.0	2.1	4.1
LSD(5%)		0.08	0.09	0.5	0.07	>0.3

[a]When calculated from the surface area of soil in the pots and assuming 1.6 kg soil/pot, 50 mg P/ kg soil is approximately equivalent to 72 kg P/ha.

In contrast, Hoepfner et al. [108] found that inoculating apple had no effect on growth in a very fertile soil without added phosphorus; and in relatively less fertile soils, inoculation had no significant beneficial effects when more than 70 kg/ha phosphorus was applied (Table XI).

Most of the studies in the above list were conducted in the nursery, and subsequent growth after transplanting either was not monitored or has not been reported. However, Barrows and Roncadori [23], Biermann and Linderman [27], Cooper [42], Cornet et al. [45], Menge et al. [146], Morandi et al. [154], and Plenchette et al. [172] showed that when inoculated and uninoculated seedlings of the same size were raised in sterile media and transplanted into soils containing a normal complement of VAM fungi, the inoculated seedlings had improved transplant survival and regrowth, and subsequently produced plants which were less variable than the uninoculated ones. As far as I am aware it has not been convincingly demonstrated why these benefits occurred. One possibility is that the mycorrhizal seedlings had been pre-inoculated with more effective endophytes than those present in the soil into which they were being transplanted. Another possibility is that as with pre-infected transplanted crops (see section II.C) the lag between transplanting and the establishment of VAM infections limited the growth of the nonmycorrhizal seedlings after transplanting. In contrast to these experiments, Snellgrove and Stribley [212] failed to detect any benefit on onion

TABLE XII. Effects of Pre-Inoculating With VAM on Peat Module Raised Onions (cv. Balstora) Transplanted to the Field With a Soil Containing 31 µg/g Bicarbonate-Extractable Phosphorus [From 212—Without Dazomet Treatments]

| | | At transplanting | | At harvest | |
| | | | | | |
Module type	VAM inoculum	Shoot fresh wt (mg)	Shoot phosphorus content (%)	Harvestable yield (t/ha—$\log_{10}$ transformed)	Root length infected (%)
Commercial blocking compost	–	550	1.22	1.59	15
Modified low phosphorus compost	–	348	0.53	1.39	24
Modified low phosphorus compost	+	363	0.58	1.60	15
				0.085 LSD(5%)	

yields from inoculation prior to transplanting into a field soil where water but not phosphorus was limiting (Table XII). Unfortunately, in order to ensure that the inoculated seedlings were well infected they used a potting mix for these seedlings which was less fertile than the commercial potting mix the controls were raised in. Consequently, at transplanting the inoculated seedlings were smaller and had much lower shoot phosphorus concentrations than the controls. The inoculated seedlings were therefore disadvantaged when transplanted to the field, and conceivably this could have affected their final yields.

C. Transplanted Crops—Seedlings Raised in Unsterile Soils

In developing countries, where labor is relatively cheap, seedlings of crops which might otherwise be field-sown can be raised in nurseries and then transplanted to the field after a previous crop has been harvested. The advantages of this are that the length of time a crop is growing in the field is reduced, as is the gap between one crop and the next, more crops can be harvested per year, and food production per unit area is increased. In these regions the soils are often particularly low in available phosphorus [22,198], but the cost of fertilizer is relatively high, which restricts its use. Crops are therefore often grown in soils containing inadequate phosphorus for maximum growth and VAM make a major contribution to their phosphorus nutrition. Simply by broadcasting 1.25 kg inoculum/m^2 of pre-selected VAM fungi over the surface of nursery beds at sowing prior to transplanting to the field, Bagyaraj and Sreeramulu [20] (Table XIII), Govinda Rao et al. [72], and Sreeramulu and Bagyaraj [213] (Table XIII) have obtained increased

TABLE XIII. Effect of Preinoculating Chilli (*Capsicum annum* L.) Transplants With *Glomus fasciculatum* and *G. albidum* on Fruit Yield and VAM Infection Level in Two Soils [From 20,213]

		Applied P (kg P/ha)		
	Inoculum	0	37.5	75
Fruit yield (kg/4.05m²)				
Site 1[a]	None	1.23	1.30	1.81
	G. fasciculum	1.33	1.56	—
	G. albidum	1.76	2.14	—
		LSD (5%):P 0.14	Inoculum 0.22	
Site 2[b]	None	0.27	0.37	0.43
	G. fasciculatum	0.40	0.52	—
	G. albidum	0.38	0.42	—
		LSD (5%):P 0.036	Inoculum 0.057	
Infection level (%)				
Site 1[a]	None	71	75	80
	G. fasciculatum	77	91	—
	G. albidum	100	100	—
		LSD (5%):P 5.89	Inoculum 9.31	
Site 2[b]	None	66	80	80
	G. fasciculatum	100	100	—
	G. albidum	89	97	—
		LSD (5%):P 5.71	Inoculum 9.03	

[a]Site 1 was at Chikkaballapur with a red sandy soil, pH 6.0, 6μg/g of $NH_4F + HCl$ extractable phosphorus. The chilli cultivar was Jwala.
[b]Site 2 was at Rattinhalli with a black clay soil, pH 7.2, and 12 μg/g of $NH_4F + HCl$ extractable phosphorus. The chilli cultivar was Byadigi.

yields of chilli and finger millet. Bagyaraj and Sreeramulu [20] and Sreeramulu and Bagyaraj [213] (Table XIII) also found that *Glomus fasciculatum* was superior at one site, while *G. albidum* Walker and Rhodes was superior at another. Unfortunately, different host cultivars were used at the two sites, and consequently it is not possible to distinguish whether these differences in fungal effectiveness were due to adaption of the fungi to soil conditions or host cultivar, both of which are known to influence responses to VAM inoculation (Table XIV).

D. Persistence of Responses to VAM Inocula

In soils naturally containing no VAM fungi or VAM fungi less effective than those which can be introduced, a response to inoculation could be expected to persist indefinitely provided the fungi in the inocula were well adapted to the soil and host, and were able to compete [227] with the

TABLE XIV. Factors Shown to Play a Differential Role in Determining the Effectiveness of Individual Inoculant VAM Fungi

  i) Soil P status [89,204,205,218].
 ii) Ability to compete with other VAM fungi [11,49,52,177,194,227].
iii) Temperature [52,203,209,221].
 iv) Adaption to soil [25,73,74,121,139,155,165,174,204,213].
  v) Adaption of fungus to host (or vice versa) [25,49,51,53,68,79,80,97,129,134,163].
 vi) Resistance to heavy metals [69,70].
vii) Soil aeration [196].
viii) Susceptibility to organisms parasitic on VAM fungi [18,132,194,207,217].
 ix) Ability to counteract the effects of pathogens of the host [18,54,55,202].

TABLE XV. Some Factors Which May Affect the Effectiveness of Individual Inoculant VAM Fungi

  i) Ability to stimulate uptake of nutrients other than P [4,43,44,97,126].
 ii) Ability to stimulate uptake of soil water [12,14,43,52,75,96,97,110,161,206,214].
iii) Ability to produce plant growth regulators [21,43].
 iv) Dependency of host on VAM formation [15,24,26,31,76,87,94,128,134,140,149, 150,171,173] .

indigenous soil microbiota. However, if a soil's natural inoculum potential was low, for example, perhaps by precropping with a nonmycorrhizal species [but see 164] such as a brassica [66] and the indigenous VAM fungi were as effective as those in the inocula, inoculation would merely produce a transitory increase in infection level. The size of the growth response would then be determined by the length of the delay between germination and the establishment of good infections in uninoculated plants compared with inoculated ones.

Hall [89] has argued that where there have been relatively recent major changes to vegetation and soils perhaps it is to be expected that the indigenous VAM fungi may no longer be those which are best suited to the new conditions. But in VAM field and pot experiments in unsterile soil it is not even necessary to have to assume that this situation preexisted, as the soil conditions and/or host species were often changed at the start of the experiments. If these changes were relatively transitory—for example, the effects of a single application of lime on pH—then responses to inoculation might also be transitory. Indeed, some pot experiments on unsterile soil do appear to demonstrate that responses to inoculation decrease with time (Fig. 4) [47,89,186,198]. However, the sizes of the responses to inoculation Hall [89] and Powell and Daniel [186] detected at early harvests were very large (Fig. 4), while at later harvests the responses to inoculation fell to much more

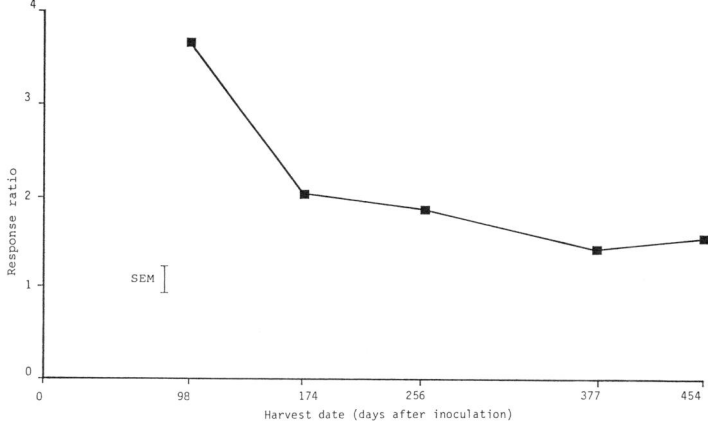

Fig. 4. VAM response ratios (means of eight different inocula and three levels of applied phosphorus—2.2, 6.6, and 19.8 mg P/pot) of Grasslands Huia white clover growing in unsterile phosphorus-deficient soil cores with five sequential harvests [from 89].

modest levels on a par with those later detected in the field [90]. Consequently, the extrapolation to the field of the fall in response to inoculation with time detected in pot experiments must be regarded as questionable. Furthermore in some of Hall's [90] (unpublished data) pastoral field experiments, responses to inoculation continued for more than 3 yr, which suggests that if there is likely to be a loss of response with time, in the field its onset is not rapid.

E. Failures of VAM Inocula to Produce Growth Responses

Within their experiments on poor soils, Rangeley et al. [188] (Table VIII) and Hall [90: experiments 1 and 2 in the second year] detected responses to inoculation only if phosphorus was applied. Indeed, in one of Rangeley et al.'s experiments, *Glomus etunicatum* Becker and Gerd. actually depressed growth unless 40 kg phosphorus/ha was also applied (Table IX). As both Hall's and Rangeley and colleagues' most effective fungi originated from relatively fertile soils, it is conceivable that their endophytes were not adapted to functioning symbiotically in soils with a very low phosphorus status. Early attempts to stimulate clover growth in the field by Hall (unpublished data) also failed because the isolate of *G. fasciculatum* used was no more effective than the indigenous fungi at a low level of available soil phosphorus. Jensen's [119] and Ross and Harper's [193] failure to detect responses in unsterile soil could also have been because the inocula were either not effective or no more effective than the indigenous ones. Indeed, in both Jensen's and Ross and Harper's studies, the inocula contained fungi derived from the experimental sites, and apparently no attempt had been made to select for effectiveness. There are, however, many reasons—other than that the inoculant fungi were no more effective than the indigenous ones—that responses to inoculant fungi may not be detected. For example, Bolan et al. [29] and Pairunan et al. [169] detected little or no response to inoculation in subterranean clover at very low levels of available soil phosphorus as well as at very high levels. Similarly, Black and Tinker [28] (Table IV) probably failed to get a response to inoculation in their plus-phosphorus treatment because this raised the soil phosphorus status to a level where VAM had no beneficial effects (Fig. 1). Newbould and Rangeley [162] also failed to get a response to inoculation on a brown earth. But they had poor nodulation of their clovers, and as nitrogen was probably the principal factor limiting growth, it would have masked any beneficial effects a more effective VAM fungus may have had on plant phosphorus uptake.

III. SELECTION OF VAM FUNGI

An excellent discussion of the selection of VAM fungi for possible use in agriculture has been published by Abbott and Robson [3], and consequently I have tended to restrict my comments to those papers which have appeared since their paper was written in 1980.

The variations in VAM fungal effectiveness may be due to differences in the ratio of hyphae a fungus produces in the soil to the amount of mycelium in the root which supports it [3,74,200]. Abbott and Robson [6] also speculated that the distribution of hyphae in the soil and a number of other factors [4] may also be important. Whatever the reasons for these differences it is known that a variety of factors can or might modify the subsequent effectiveness of a VAM fungus and may have to be taken into account when selecting VAM fungi for a specific purpose. For example, pH can affect the germination of VAM fungal spores, hyphal growth, and the effectiveness of endophytes [7,58,79,81,106,122,135,208,219,224]. Other factors known to, or which might, have similar effects are listed in Table XIV and XV along with references to pertinent papers.

Govinda Rao et al. [72], Gianinazzi-Pearson et al. [67], Hall [89], Powell [180], and Schubert and Hayman [204], for example, have outlined procedures for comparing the effectiveness of VAM fungi. Govinda Rao and co-workers' [72] technique was designed specifically to select fungi for the pre-inoculation of transplanted crops raised in unsterile soils (section II.C).

Seedlings were first raised in unsterile nursery soils containing a normal complement of indigenous mycorrhizal fungi to which either VAM cultures on guinea grass (*Panicum maximum* Jacq.) or a similar amount of an uninfected grass root and soil mixture had been added. Once infections had developed, the seedlings were transplanted to experimental field plots and plant growth monitored. This technique follows precisely the procedures used by farmers in India and therefore cannot be criticized.

The selection of fungi using pre-inoculation transplant techniques, however, is probably of little value for normally field-sown crops and pastures as the ability of the fungi to survive in inocula suitable for field use has not been taken into account—a criticism which can also be made of the experiments of Gianinazzi-Pearson et al. [67], Hall [89], and Schubert and Hayman [204]. These experiments can also be faulted because fungi were screened in greenhouse pot experiments and consequently all the criticisms of pot experiments outlined in section I apply. Unfortunately Schubert and Hayman [204] also compared VAM effectiveness in sterile soil, which ignores the ability of the fungi to function with a competing soil flora. Gianinazzi-Pearson and co-workers' [67] techniques can also be criticized because they used sievings of field soils as the VAM inocula instead of pure cultures. They may therefore have estimated the confounded beneficial effects of the VAM fungi and deleterious components in the inocula [41,89,199]. Most of these faults are, however, avoided in the technique devised by Powell [180]. In this, VAM fungi were first isolated from unsterile soil by baiting [85,91] with sterile perennial ryegrass (*Lolium perenne* L.) seedlings. The seedlings were then transplanted into sterile soil, and the ability of the endophytes to stimulate growth was compared by weighing harvested herbage. Those fungi which stimulated growth most were then evaluated in several unsterile soils in a shadehouse and in the field. Unfortunately his field comparisons involved inappropriate pre-inoculated transplant techniques. However, in his shadehouse study, soils infested with *Glomus fasciculatum,* or the initial fungal selections or an uninfested control soil were each made into pellets containing two white clover seeds. These were then sown onto a number of unsterile soils in pots; and after 13 wk, plant growth was measured and the effects of the VAM fungi compared. This is an approach which has much to recommend it. The inoculant fungi were introduced into the soil in soil pellets which may have some practical value (section IV), and once the seedling has passed through the pellet the inoculant fungus has to compete with the indigenous VAM fungi and soil flora and fauna before establishing an infection.

It is conceivable that some fungi with potentially desirable characteristics such as a superior ability to persist in the soil [3] may have been discarded during Powell's initial selection, but he has argued (personal communication) that this is a risk one might have to take to reduce the number of fungi being further assessed to a manageable level. The effectiveness of VAM fungi can vary with soil phosphorus level (Table II), and hence screening of VAM fungi should be conducted with several levels of applied phosphorus [e.g., 72,89] instead of the one level Powell used [180]. Powell's experiments could also have been extended to screen for ability to counteract the deterimental effects of pathogens and various other factors listed in Tables XIV and XV. Obviously experiments incorporating these additional factors would be very large and quite unmanageable unless confounded designs or resolvable balanced incomplete blocking designs were employed [40].

Daft and Hogarth [49] showed that inocula containing more than one endophyte gave more consistent results than those containing a single species. They therefore suggested mixtures of VAM fungi be used for field inoculations. But there would be no value in including a species in a mixed inoculum which had no redeeming features, and therefore screening would still have to be carried out to eliminate these. Moreover, as competition between VAM can occur [11,177,227] careful consideration would have to be given to the composition of mixed inocula to avoid the inclusion of one which might oust the others.

IV. INOCULA AND INOCULUM PRODUCTION

A number of pot investigations have shown that inoculant VAM fungi can spread through VAM fungus infested soil at 300 to >1,000 mm per year [11,159,177]. Also Jakobsen [115] has detected a rate of spread of 300 mm in 96 days through fumigated soil in the field. If similar rates of spread for inoculant VAM fungi can be demonstrated through unsterilized field soils, then it should be possible to reduce the amount of pelleted whole soil inoculum to just a few tens of kilograms per hectare [90]. However, lighter and possibly more potent inocula such as mass-produced spores [152] or those produced using the nutrient film technique [59,60,145,158], root organ cultures [37], in expanded clay [56] or peat cultures (Mosse, personal communication) [226] might prove more attractive than whole soil for field use.

In most field experiments conducted to date the inoculum was whole soil containing VAM spores, hyphae, and infected roots (see ref 145 for a review of pot culture methods for the production of inocula). In some cases it was pelletized using clay binding agents [90,95,103] which made it more convenient to apply, and by placing it in the seeding furrows, i.e., in the immediate vicinity of the seed, made it more effective than broadcast inoculum [104]. VAM have also been successfully introduced into soils by inoculating seed or seedlings with fresh or lyophilized VAM fungal spores [64,100], fresh roots [e.g., 84,89], dried whole inocula [137], lyophilized roots [48], and by fluid-drilling soil sievings or homogenized spores, hyphae, and roots [57,63,104].

The first roots of VAM hosts which have relatively small seeds containing limited quantities of phosphorus are quickly infected by VAM fungi [e.g., 210]. However, species with relatively large seeds can contain considerable stores of phosphorus, which seems to make their seedling roots uninfectable and hence inoculation is more successful if the inoculum is placed a few centimeters below the seed (e.g., soybean, corn, sorghum [114]; peach [141]; citrus—Menge, personal communication).

Some VAM fungi can be adversely affected by other microorganisms (see Table XIV: viii for references), and obviously it is important that inocula should be free of such hyperparasites. But perhaps even more importantly it should be ensured that inocula produced for commercial purposes are free of oragnisms pathogenic to the host; and to achieve this it may be necessary to go to considerable lengths [145,150]. Commercially produced inocula must also maintain their integrity and not be subject to drift in their effectiveness [47] or to contamination with other, perhaps less desirable VAM fungi [3,85].

V. CONCLUSIONS AND FUTURE RESEARCH

Research funding bodies are often populated by individuals who are not scientists themselves but who have specialist skills in other directions. To extract money from these bodies against competition it is sometimes necessary to make a case outlining the most favorable outcome of a research program. Unfortunately, when this is written by an overenthusiastic researcher or similarly modified by a well-meaning head of a department, it can lead to more optimism than is really justified. I can think of one example at least where this has occurred and which resulted in a suggestion that farmers would be using VAM inocula by the end of the decade—a prediction made in 1977! The result was predictable: a loss of confidence in the research, a withdrawal of funding, and the collapse of the research program. Of the papers which have been published on VAM, those which deal with field applications are very much in the minority. Far more, however, expound the potentials for exploiting the symbiosis either in their introductions or discussions, and I believe that this too along with spectacular responses in greenhouse pot experiments has added to perhaps unjustified optimism.

A. Field-Sown Crops and Pastures

Whether VAM technology will ever be used in the production of field-sown crops and pastures will depend on the economics of employing it as compared with, for example, stimulating growth with fertilizers. But in past field experiments (see section II.A) the primary consideration was to gauge the effects inoculant fungi had on growth. Whether the method of inoculation was economically justifiable or had any agronomic value was of secondary interest. All that really mattered was that the host plant had as good a chance as possible of picking up the inoculant fungus and that the experiment did not founder for want of inoculum potential. The levels of inocula applied in the experiments were therefore very high, generally in the range 0.8 t/ha [111] to 25 t/ha [9] although 100 t/ha was used by Lambert and Cole [138], all of which were well beyond what could be considered practical. Consequently, before an economic appraisal of the value of inoculating field-sown crops and pastures with VAM inocula can be made, more information is needed on the minimum quantity of inocula that has to be applied and the most cost-effective way of producing it. Additional work is also required on identifying very effective endophytes, the soil/host combinations to which they are best suited, and how long responses to inoculation are likely to last. Of these, I think the most important is our inability to produce large quantities of cheap, reliable inocula, for without this, research on the effects of inoculating field-sown crops and pastures becomes little more than a stimulating academic exercise. Dehne and Backhaus's [56] technique of growing inocula inside expanded clay in pot cultures and the peat cultures of Warner et al. [226] appear to be suited to commercial use. However, I believe that the production of large quantities of inocula in pure axenic culture [107] is the only way that inocula will become cheap enough to be employed as rhizobial inocula are currently used in agriculture.

B. Transplanted Crops

From the work of Bagyaraj and co-workers (see section II.C) and, for example, the research on citrus (see section III.B) there can be little doubt that VAM inocula can produce worthwhile and economic growth responses for those transplanted species which are raised in unsterile or sterile soil in nurseries, are transplanted into relatively phosphorus-deficient soils, and where the cost of phosphatic fertilizers is a severe economic constraint. At the other extreme, where VAM have no detectable effects other than with the phosphorus nutrition of a crop and the cost of phosphatic fertilizer is negligible compared with the value of the crop [212], VAM can be ignored.

Between these two ends of the scale—where either the cost of fertilizer phosphorus is an important consideration or where VAM have some beneficial effect, other than with the phosphorus nutrition of a crop, which cannot otherwise be achieved cheaply—is an area where there is insufficient information at present to predict the future importance of VAM. Clearly this is an area which requires an additional research input.

ACKNOWLEDGMENTS

I wish to thank Invermay's long-suffering typists and my colleagues for their helpful comments during the preparation of this review. I am also grateful to the Miss E. L. Hallaby Trust for financial assistance. It should be noted that the reference list is not exhaustive, and the absence of a paper does not necessarily indicate that it is inferior to another which is cited.

I am grateful to those who allowed me to use their copyrighted published material, including especially the following: Academic Press Inc. (Tables V, VI, VIII), Agricultural Institute of Canada (Table X), CSIRO Editorial and Publishing Unit (Table VII), Cambridge University Press (Table IX; Fig. 2, 4), DSIR Science Information Publishing Centre (Table III; Fig. 1), Journal of the American Society for Horticultural Science (Table XI), Zeitschrift fur Angewandte Botanik (part of section II.A.2) [92], Journal of Nematology (Table I), Martinus Nijhoff (Tables II, XII), Nature (Table IV), and the Soil Science Society of America (Fig. 3).

REFERENCES

1. Abbott LK, Robson AD: Growth stimulation of subterranean clover with vesicular-arbuscular mycorrhiza. Aust J Agric Res 28:639, 1977.
2. Abbott LK, Robson AD: Growth of subterranean clover in relation to the formation of endomycorrhizas by introduced and indigenous fungi in a field soil. New Phytol 81:575, 1978.
3. Abbott LK, Robson AD: The role of vesicular-arbuscular mycorrhizal fungi in agriculture and the selection of fungi for inoculation. Aust J Agric Res 33:389, 1982.
4. Abbott LK, Robson AD: The effect of mycorrhizae on plant growth. In Powell CLI, Bagyaraj DJ (eds): "VA Mycorrhizas." Boca Ration: CRC Press, 1984, p 113.
5. Abbott LK, Robson AD: Colonization of the root system of subterranean clover by three species of vesicular-arbuscular mycorrhizal fungi. New Phytol 96:275, 1984.
6. Abbott LK, Robson AD: Formation of external hyphae in soil by four species of vesicular-arbuscular mycorrhizal fungi. New Phytol 99:245, 1985.
7. Abbott LK, Robson AD: The effect of soil pH on the formation of VA mycorrhizas by two species of Glomus. Aust J Soil Res 23:253, 1985.
8. Abbott LK, Robson AD, Deboer G: The effect of phosphorus on the formation of hyphae in soil by the vesicular-arbuscular mycorrhizal fungus Glomus fasciculatum. New Phytol 97:437, 1984.

9. Abbott LK, Robson AD, Hall IR: Introduction of vesicular-arbuscular mycorrhizal fungi into agricultural soils. Aust J Agric Res 34:741, 1983.

10. Aldon EF: Endomycorrhizae enhance shrub growth and survival on mine spoils. In Wright RA (ed): "The Reclamation of Disturbed Arid Lands." Albuquerque: University of New Mexico Press, 1978, p 174.

11. Aldwell FEB, Hall IR: Monitoring the spread of *Glomus mosseae* through soil infested with *Acaulospora laevis* using serological and morphological techniques. Trans Br Mycol Soc 87:131, 1986.

12. Allen MF: Influence of vesicular-arbuscular mycorrhizae on water movement through *Bouteloua gracilis* (H.B.K.) Lag. ex Steud. New Phytol 91:191, 1982.

13. Ames RN, Linderman RG: The growth of easter lily *Lillium longiflorum* as influenced by VAM fungi, *Fusarium oxysporum,* and fertility level. Can J Bot 56:2773, 1978.

14. Auge RM, Schekel KA, Wample RL: Greater leaf conductance of well watered VA mycorrhizal rose plants is not related to phosphorus nutrition. New Phytol 103:107, 1986.

15. Azcon R, Ocampo JA: Factors affecting the vesicular-arbuscular infection and mycorrhizal dependancy of thirteen wheat cultivars. New Phytol 87:677, 1981.

16. Azcon-Aguilar C, Barea JM: Field inoculation of *Medicago sativa* with vesicular-arbuscular mycorrhiza and *Rhizobium n:eliloti* in phosphate fixing agricultural soil. Soil Biol Biochem 13:19, 1981.

17. Azcon-G de Aguilar C, Azcon R, Barea JM: Endomycorrhizal fungi and *Rhizobium* as biological fertilisers for *Medicago sativa* in normal cultivation. Nature 279:325, 1979.

18. Bagyaraj DJ: Biological interactions with VA mycorrhizal fungi. In Powell CLI, Bagyaraj DJ (eds): "VA Mycorrhizas." Boca Raton: CRC Press, 1984, p 131.

19. Bagyaraj DJ, Manjunath A, Patil RB: Interaction between a vesicular-arbuscular mycorrhiza and *Rhizobium* and their effects on soybean in the field. New Phytol 82:141, 1979.

20. Bagyaraj DJ, Sreeramulu KR: Preinoculation with VA mycorrhiza improves growth and yield of chilli transplanted in the field and saves phosphatic fertilizer. Pl Soil 69:375, 1982.

21. Barea JM, Azcon-Aguilar C: Production of plant growth-regulating substances by the vesicular-arbuscular mycorrhizal fungus *Glomus mosseae*. Appl Environ Microbiol 43:810, 1982.

22. Barnard RO, Folscher WJ: Nutrient interactions in acid tropical soil. Trop Agric (Trinidad) 57:333, 1980.

23. Barrows JB, Roncadori RW: Endomycorrhizal synthesis by *Gigaspora margarita* in poinsettia. Mycologia 69:1173, 1977.

24. Bertheau Y, Gianinazzi-Pearson V, Gianinazzi S: Développement et expression de l'association endomycorhizienne chez le ble. I—Mise en évidence d'un effect variétal. Ann Amelior Plant 30:67, 1980.

25. Bethlenfalvay GJ, Ulrich JM, Brown MS: Plant responses to mycorrhizal fungi: Host, endophyte and soil effects. Soil Sci Soc Am J 49:1164, 1982.

26. Bhattarai ID, Mishra RR: Study on the vesicular-arbuscular mycorrhiza of three cultivars of potato (*Solanum tuberosum* L.) Pl Soil 79:299, 1984.

27. Biermann BJ, Linderman RG: Increased geranium growth using pretransplant inoculation with a mycorrhizal fungus. J Am Soc Hort Sci 108:972, 1983.

28. Black RLB, Tinker PB: Interactious between effects of vesicular-arbuscular mycorrhiza and fertiliser phosphorus on yield of potatoes in the field. Nature 267:510, 1977.

29. Bolan NS, Robson AD, Barrow NJ: Plant and soil factors including mycorrhizal infection causing sigmoidal response of plants to applied phosphorus. Pl Soil 73:187, 1983.

30. Bowen GD, Rovira AD: The influence of microorganisms on root growth and metabolism. In Wittington WJ (ed): "Root Growth." London: Butterworths, 1969, p 170.

31. Brown RW, Schultz RC, Kormanik PP: Response of vesicular-arbuscular endomycorrhizal sweetgum seedlings to three nitrogen fertilizers. For Sci 27:413, 1981.

32. Busse MD, Ellis JR: Vesicular-arbuscular mycorrhizal *(Glomus fasciculatum)* influence on soybean drought tolerance in high phosphorus soil. Can J Bot 63:2290, 1985.

33. Buwalda JG, Stribley DP, Tinker PB: Effects of vesicular-arbuscular mycorrhizal infection in first, second and third cereal crops. J Agric Sci 105:631, 1985.

34. Buwalda JG, Stribley DP, Tinker PB: Vesicular-arbuscular mycorrhizae of winter and spring cereals. J Agric Sci 105:649, 1985.

35. Caron M, Fortin JA, Richard DC: Effect of *Glomus intraradices* on infection by *Fusarium oxysporum* f.sp. *radicis lycopersici* in tomatoes over a 12-week period. Can J Bot 64:552, 1986.

36. Caron M, Fortin JA, Richard C: Effect of phosphorus concentration and *Glomus intraradices* on fusarium crown and root rot of tomatoes. Phytopathology 76:942, 1986.

37. Carr JP, Hinkley MA, Jones MG: Mycorrhizal inoculum production. In "Rothamsted Experimental Station Report for 1982, Part 1." Harpenden: Rothamsted Experimental Station, 1983, p 166.

38. Chulan-Hashim A, Ragu P: Growth response of *Theobroma cacao* L. seedlings to inoculation with vesicular-arbuscular mycorrhizal fungi. Pl Soil 96:279, 1986.

39. Clarke C, Mosse B: Plant Growth responses to VAM. XII. Field inoculation responses of barley at two soil P levels. New Phytol 87:695, 1981.

40. Cochran WG, Cox GM: "Experimental Design," Ed 2. London: Wiley, 1957.

41. Cooper KM: Growth responses to the formation of endotrophic mycorrhizas in *Solanum, Leptospermum,* and New Zealand ferns. In Sanders FE, Mosse B, Tinker PB (eds): "Endomycorrhizas." London: Academic Press, 1975, p 391.

42. Cooper KM: The role of VA mycorrhizas in the development of a new commercial crop—tamarillo—in New Zealand. Proc 5th North Am Conf Mycorrhizae, Univ Laval Quebec:54(Abstr), 1981.

43. Cooper KM: Physiology of VA mycorrhizal associations. In Powell CLI, Bagyaraj DJ (eds): "VA Mycorrhiza." Boca Raton: CRC Press, 1984, p 155.

44. Cooper KM, Tinker PB: Translocation and transfer of nutrients in vesicular-arbuscular mycorrhizas II. Uptake and translocation of phosphorus, zinc and sulphur. New Phytol 81:43, 1978.

45. Cornet F, Diem HG, Dommergues YR: Effect de l'inoculation avec *Glomus mosseae* sur la croissance d'*Acacia holosericea* an pepiniére et apres transplantation sur le terrain. In Gianinazzi S, Gianinazzi-Pearson V, Trouvelot A (eds): "Mycorrhiza, an Integral Part of Plants: Biology and Prospects for Their Use." Paris: Institut National de la Recherche Agronomique, 1982, p 287.

46. Crews CL, Johnson CR, Joiner JN: Benefits of mycorrhizae on growth and development of three woody ornamentals. Hort Sci 13:429, 1978.

47. Crush JR: Changes in effectiveness of soil endomycorrhizal fungal populations during pasture development. NZ J Agric Res 21:683, 1978.

48. Crush JR, Pattison AC: Preliminary results on the production of vesicular-arbuscular

mycorrhizal inoculum by freeze drying. In Sanders FE, Mosse B, Tinker PB (eds): "Endomycorrhizas." London: Academic Press, 1975, p 483.

49. Daft MJ, Hogarth BG: Competitive interactions amongst four species of *Glomus* on maize and onion. Trans Br Mycol Soc 80:339, 1983.

50. Daniels BA, McCool PM, Menge JA: Comparative inoculum potential of spores of six vesicular-arbuscular mycorrhizal fungi. New Phytol 89:385, 1981.

51. Daniels BA, Menge JA: Evaluation of the commercial potential of the vesicular-arbuscular mycorrhizal fungus, *Glomus epigaeus*. New Phytol 87:345, 1981.

52. Daniels-Hetrick BA: Ecology of VA mycorrhizal fungi. In Powell CLI, Bagyaraj DJ (eds): "VA Mycorrhizas." Boca Raton: CRC Press, 1984, p 35.

53. Daniels-Hetrick BA, Bloom J: The influence of host plant on production and colonization ability of vesicular-arbuscular mycorrhizal spores. Mycologia 78:32, 1986.

54. Davis RM, Menge JA: *Phytophthora parasitica* inoculation and intensity of vesicular-arbuscular mycorrhizae in citrus. New Phytol 87:705, 1981.

55. Dehne HW: Interaction between vesicular-arbuscular mycorrhizal fungi and plant pathogens. Phytopathology 72:1115, 1982.

56. Dehne HW, Backhaus GF: The use of vesicular-arbuscular mycorrhizal fungi in plant production 1. Inoculum production. Pflanzenkr Pflanzenschutz 93:415, 1986.

57. Diem HG, Jung G, Mugnier J, Ganry F, Dommergues YR: Alginate-entrapped *Glomus mosseae* as an inoculant. In Fortin JA, Furlan V (eds): "Program and Abstracts of the Fifth North American Conference on Mycorrhizas." Quebec: Universite Laval, 1981, p 23.

58. Dudeck AE, Schenck NC, Peacock CH: Influence of mycorrhizae on the growth of bahiagrass and centipedegrass. Soil Crop Sci Soc Fl Proc 43:137, 1984.

59. Elmes RP, Hepper CM, Hayman DS, O'Shea J: The use of vesicular-arbuscular mycorrhizal roots grown by the nutrient film technique as inoculum for field sites. Ann Appl Biol 104:437, 1984.

60. Elmes RP, Mosse B: Vesicular-arbuscular endomycorrhizal inoculum production. II. Experiments with maize *(Zea mays)* and other hosts in nutrient flow culture. Can J Bot 62:1531, 1984.

61. Fairweather JV, Parbery DG: Effects of four vesicular-arbuscular mycorrhizal fungi on growth of tomato. Trans Br Mycol Soc 79:151, 1982.

62. Furlan V, Fortin JA, Plenchette C: Effects of different vesicular-arbuscular mycorrhizal fungi on growth of *Fraxinus americana*. Can J For Res 13:589, 1983.

63. Ganry YF, Diem HG, Dommergues YR: Effect of inoculation with *Glomus mosseae* on nitrogen fixation by field grown soy beans. Pl Soil 68:321, 1982.

64. Gaunt RE: Inoculation of vesicular-arbuscular mycorrhizal fungi on onion and tomato seeds. NZ J Bot 16:69, 1978.

65. Geddeda YI, Trappe JM, Stebbins RL: Effects of vesicular-arbuscular mycorrhizae and phosphorus on apple seedlings. J Am Soc Hort Sci 109:24, 1984.

66. Gerdemann JW: Vesicular-arbuscular mycorrhizae. In Torrey JG, Clarkson DT (eds): "The Development and Function of Roots." London: Academic Press, 1975, p 575.

67. Gianinazzi-Pearson V, Gianinazzi S, Trouvelot A: Evaluation of the infectivity and effectiveness of indigenous vesicular-arbuscular fungal populations in some agricultural soils in Burgundy. Can J Bot 63:1521, 1985.

68. Giovannetti M, Hepper CM: Vesicular-arbuscular mycorrhizal infection in *Hedysarum coronarium* and *Onobrychis viciaefolia*: Host-endophyte specificity. Soil Biol Biochem 17:899, 1985.

69. Gildon A, Tinker PB: A heavy metal tolerant strain of mycorrhizal fungus. Trans Br Mycol Soc 77:648, 1981.

70. Gildon A, Tinker PB: Interactions of vesicular-arbuscular mycorrhizal infections and heavy metals in plants. I. The effects of heavy metals on the development of vesicular-arbuscular mycorrhizas. New Phytol 95:247, 1983.

71. Gould AB, Liberta AE: Effects of topsoil storage during surface mining on the viability of vesicular-arbuscular mycorrhiza. Mycologia 73:914, 1981.

72. Govinda Rao YS, Bagyaraj DJ, Rai PV: Selection of an efficient vesicular-arbuscular mycorrhizal fungus for finger millet. 2. Screening under field conditions. Zentralbl Mikrobiol 138:415, 1983.

73. Grace L: Changes in endophyte efficiency in different soils. In "Rothamsted Experimental Station Report for 1982, Part 1." Harpenden: Rothamsted Experimental Station, 1983, p 218.

74. Graham JH, Linderman RG, Menge JA: Development of external hyphae by different isolates of mycorrhizal Glomus spp. in relation to root colonization and growth of troyer citrange. New Phytol 91:183, 1982.

75. Graham JH, Syvertsen JP: Influence of vesicular-arbuscular mycorrhiza on the hydraulic conductivity of roots of two citrus root stocks. New Phytol 97:277, 1984.

76. Graham JH, Syvertsen JP: Host determinants of mycorrhizal dependency of citrus rootstock seedlings. New Phytol 101:667, 1985.

77. Graham JH, Timmer LW: Vesicular-arbuscular mycorrhizal development and growth response of rough lemon in soil and soilless media: Effect of phosphorus source. J Am Soc Hort Sci 109:118, 1984.

78. Grandison GS, Cooper KM: Interaction of Vesicular-arbuscular mycorrhizae and cultivars of alfalfa susceptible and resistant to *Meloidogyne hapla*. J Nematol 18:141, 1986.

79. Graw D: The influence of soil pH on the efficiency of vesicular-arbuscular mycorrhiza. New Phytol 82:687, 1979.

80. Graw D, Moawad M, Rehm S: Untersuchungen zur Wirts- und Wirkungsspezifitat der VA-Mykorrhiza. Acker Pflanzenbau 148:85, 1979.

81. Green NE, Graham SO, Schenck NC: The influence of pH on the germination of vesicular-arbuscular mycorrhizal spores. Mycologia 68:929, 1976.

82. Haas JH, Bar-Yosef B, Krikun J, Barak R, Markovitz T, Kramer S. Vesicular-arbuscular mycorrhizal fungus infestation and phosphorus fertigation to overcome pepper stunting after methyl bromide fumigation. Agron J 79:905, 1987.

83. Hall IR: Endomycorrhizas of *Metrosideos umbellata* and *Weinmannia racemosa*. NZ J Bot 13:463, 1975.

84. Hall IR: Response of *Coprosma robusta* to different forms of endomycorrhizal inoculum. Trans Br Mycol Soc 67:409, 1976.

85. Hall IR: Species and mycorrhizal infections of New Zealand Endogonaceae. Trans Br Mycol Soc 68:341, 1977.

86. Hall IR: Effect of endomycorrhizas on the competitive ability of white clover. NZ J Agric Res 21:509, 1978.

87. Hall IR: Effect of vesicular-arbuscular mycorrhizas on two varieties of maize and one of sweetcorn. NZ J Agric Res 21:517, 1978b.

88. Hall IR: Growth of *Lotus pedunculatus* Cav. in an eroded soil containing soil pellets infested with endomycorrhizal fungi. NZ J Agric Res 23:103, 1980.

89. Hall IR: Effect of inoculant endomycorrhizal fungi on white clover growth in soil cores. J Agric Sci 102:719, 1984.

90. Hall IR: Field trials assessing the effect of inoculating agricultural soils with endomy-corrhizal fungi. J Agric Sci 102:725, 1984.
91. Hall IR: Taxonomy of VA mycorrhizal fungi. In Powell CLI, Bagyaraj DJ (eds): ''VA Mycorrhizas.'' Boca Raton: CRC Press, 1984, p 57.
92. Hall IR: A review of VA mycorrhizal growth responses to pastures. Angew Bot 61:127, 1987.
93. Hall IR, Armstrong P: The effect of vesicular-arbuscular mycorrhizas on growth of white clover, lotus and ryegrass in some eroded soils. NZ J Agric Res 22:479, 1979.
94. Hall IR, Johnstone PD, Dolby R: Interactions between endomycorrhizas and soil nitrogen and phosphorus on the growth of ryegrass. New Phytol 97:447, 1984.
95. Hall IR, Kelson A: An improved technique for the production of endomycorrhizal infested soil pellets. NZ J Agric Res 24:221, 1981.
96. Hardie K, Leyton L: The influence of vesicular-arbuscular mycorrhiza on growth and water relations of red clover. I. In phosphate deficient soil. New Phytol 89:599, 1981.
97. Harley JL, Smith SE: ''Mycorrhizal Symbiosis.'' London: Academic Press, 1983.
98. Harris D, Pacovsky, Paul EA: Carbon economy of soybean—*Rhizobium-Glomus* associations. New Phytol 101:427, 1985.
99. Hashim-Chulan A: ''The Vesicular-Arbuscular (VA) Endophyte and Its Implications to Malaysian Agriculture.'' PhD Thesis, Universiti Kebangsaan Malaysia, Bangi, Malaysia, 1986.
100. Hattingh MJ, Gerdemann JW: Inoculation of Brazilian sour orange seed with endomy-corrhizal fungus. Phytopathology 65:1013, 1975.
101. Hayman DS: Mycorrhizal effects on white clover in relation to hill land improvement. Agric Res Counc Res Rev 3:82, 1977.
102. Hayman DS: White clover in Welsh hill grasslands. In ''Rothamsted Experimental Station Report for 1982, Part 1.'' Harpenden: Rothamsted Experimental Station, 1983, p 215.
103. Hayman DS: Improved establishment of white clover in hill grasslands by inoculation with mycorrhizal fungi. In Thomson DJ (ed): ''Forage Legumes.'' Occasional Sympo-sium 16. Hurley: British Grassland Society, 1984, p 44.
104. Hayman DS, Morris EJ, Page RJ: Methods for inoculating field crops with mycorrhizal fungi. Ann Appl Biol 99:247, 1981.
105. Hayman DS, Mosse: Improved growth of white clover in hill grasslands by mycorrhizal inoculation. Ann Appl Biol 93:141, 1979.
106. Hepper CM: Regulation of spore germination of the vesicular-arbuscular mycorrhizal fungus *Acaulospora laevis* by soil pH. Trans Br Mycol Soc 83:154, 1984.
107. Hepper CM: Isolation and culture of VA mycorrhizal (VAM) fungi. In Powell CLI, Bagyaraj DJ (eds): ''VA Mycorrhiza.'' Boca Raton: CRC Press, 1984, p 95.
108. Hoepfner EF, Koch BL, Covey RP: Enhancement of growth and phosphorus concentra-tions in apple seedlings by vesicular-arbuscular mycorrhizae. J Am Soc Hort Sci 108:207, 1983.
109. Howeler RH, Cadavid LF, Burckhardt E: Response of cassava to VA mycorrhizal inoculation and phosphorus application in greenhouse and field experiments. Pl Soil 69:327, 1982.
110. Huang RS, Smith WK, Yost RS: Influence of vesicular-arbuscular mycorrhiza on growth, water relations, and leaf orientation in *Leucaena leucocephala* (Lam.) de Wit. New Phytol 99:229, 1985.
111. Islam R, Ayanaba A: Effect of seed inoculation and preinfecting cowpea *(Vigna*

unguiculata) with *Glomus mosseae* on growth and seed yields of the plants under field conditions. Pl Soil 61:341, 1981.

112. Islam R, Ayanaba A, Sanders FE: Respose to cowpea *(Vigna unguiculata)* to inoculation with VA mycorrhizal fungi and to rock phosphate fertilization in some unsterilized Nigerian soils. Pl Soil 54:107, 1980.

113. Iqbal SH, Qureshi KS: The effect of VAM associations on growth of sunflower *(Helianthus annuus* L.) under field conditions. Biologia 23:189, 1977.

114. Jackson EJ, Franklin RE, Miller RH: Effects of vesicular-arbuscular mycorrhizae on growth and phosphorus content of three agronomic crops. Soil Sci Soc Am Proc 36:64, 1972.

115. Jakobsen I: Vesicular-arbuscular mycorrhiza in field-grown crops. II. Effect of inoculation growth and nutrient uptake in barley at two phosphorus levels in fumigated soil. New Phytol 94:595, 1983.

116. Jakobsen I, Andersen AJ: Vesicular-arbuscular mycorrhiza and growth in barley: Effects of irradiation and heating of soil. Soil Biol Biochem 14:171, 1982.

117. Jakobsen I, Jensen A: Influence of vesicular-arbuscular mycorrhiza and straw mulch on growth of barley. Pl Soil 62:157, 1981.

118. Jensen A: Influence of four vesicular-arbuscular mycorrhizal fungi on nutrient uptake and growth in barley. New Phytol 90:45, 1982.

119. Jensen A: The effect of indigenous vesicular-arbuscular mycorrhizal fungi on nutrient uptake and growth of barley in two Danish soils. Pl Soil 70:155, 1983.

120. Johnson CR, Joiner JN, Crews CE: Effects of N, K, Mg on growth and leaf nutrient composition of three container grown woody ornamentals inoculated with mycorrhizae. J Am Soc Hort Sci 105:286, 1980.

121. Johnson PN: Mycorrhizal endogonaceae in a New Zealand forest. New Phytol 78:161, 1977.

122. Karagiannidis N, Khanaqa A, Moawad M: Influence of soil pH on the efficiency of vesicular-arbuscular mycorrhizas. In Fortin JA, Furlan V (eds): "Program and Abstracts of the Fifth North American Conference on Mycorrhizas." Quebec: Universite Laval, 1981, p 18.

123. Khan AG: The effect of vesicular-arbuscular mycorrhizal associations on growth of cereals. I. Effects of maize growth. New Phytol 71:613, 1972.

124. Khan AG: The effect of VAM associations on the growth of cereals. II. Effects on wheat growth. Ann Appl Biol 80:27, 1975.

125. Khan AG: Growth responses of endomycorrhizal onions in unsterilized coal waste. New Phytol 87:363, 1981.

126. Killham K: Vesicular-arbuscular mycorrhizal mediation in trace and minor element uptake in perennial grasses: Relation to livestock herbage. In Fitter AH, Atkinson DA, Read DJ, Usher MB (eds): "Ecological Interactions in Soil: Plants, Microbes and Animals." Oxford: Blackwell, 1985, p 225.

127. Kormanik PP: Effect of phosphorus and vesicular-arbuscular mycorrhizae on growth and leaf retention of black walnut seedlings. Can J For Res 15:688, 1985.

128. Kormanik PP, Bryan WC, Schultz RC: Effects of three vesicular-arbuscular mycorrhizal fungi on sweetgum seedlings from nine mother trees. For Sci 27:327, 1981.

129. Kormanik PP, Schultz RC, Bryan WC: The influence of vesicular-arbuscular mycorrhizae on the growth and development of eight hardwood tree species. For Sci 28:531, 1982.

130. Krikun J, Haas JH, Bar-Yosef B. Use of VA mycorrhizal fungus inoculum in soils and semi-arid climates: A field study with bell pepper and transplants. Z Angew Bot 61:97, 1987.

131. Krishna KR, Bagyaraj DJ: Growth and nutrient uptake of peanut inoculated with the mycorrhizal fungus *Glomus fasciculatum* compared with non-inoculated ones. Pl Soil 77:405, 1984.

132. Krishna KR, Balakrishna AN, Bagyaraj DJ: Interaction between a vesicular-arbuscular mycorrhizal fungus and *Streptomyces cinnamomeous* and their effects on finger millet. New Phytol 92:401, 1982.

133. Krishna KR, Dart PJ: Effect of mycorrhizal inoculation and soluble phosphorus fertilizer on growth and phosphorus uptake of pearl millet. Pl Soil 81:247, 1984.

134. Krishna KR, Shetty KG, Dart PJ, Andrews DJ: Genotype dependent variation in mycorrhizal colonization and response to inoculation of pearl millet. Pl Soil 86:113, 1985.

135. Kucey RMN, Diab GES: Effects of lime, phosphorus, and addition of vesicular-arbuscular (VA) mycorrhizal fungi on indigenous VA fungi and on growth of alfalfa in a moderately acidic soil. New Phytol 98:481, 1984.

136. Kucey RMN, Paul EA: Vesicular-arbuscular mycorrhizal spore population in various Saskatchewan soils and the effect of inoculation with *Glomus mosseae* on faba bean growth in greenhouse and field trials. Can J Soil Sci 63:87, 1983.

137. Kuo CG, Huang RS: Effect of vesicular-arbuscular mycorrhizae on the growth and yield of rice-stubble cultured soybeans. Pl Soil 64:325, 1982.

138. Lambert DH, Cole H, Jr: Effects of mycorrhiza on establishment and performance of forage species in mine spoil. Agron J 72:257, 1980.

139. Lambert DH, Cole H, Jr, Baker DE: Adaptation of vesicular-arbuscular mycorrhizae to edaphic factors. New Phytol 85:513, 1980.

140. Lambert DH, Cole H, Jr, Baker DE: Variation in the response of alfalfa clones and cultivars to mycorrhizae and phosphorus. Crop Sci 20:615, 1980.

141. LaRue JH, McClellan WD, Peacock WI: Mycorrhizal fungi and peach nursery nutrition. Calif Agric 29:6, 1975.

142. Lopes AS, Wollum AG: Comparative effectives of methylbromide, propylene oxide, and autoclave sterilization on specific soil chemical characteristics. Turrialba 26:351, 1976.

143. Maronek DM, Hendrix JW, Kiernan J: Differential growth response to the mycorrhizal fungus *Glomus fasciculatus* of southern mognolia and Bar Harbor juniper grown in containers in composted hardwood bark-shale. J Am Soc Hort Sci 105:206, 1980.

144. Menge JA: Effect of soil fumigants and fungicides on vesicular fungi. Phytopathology 72:1125, 1982.

145. Menge JA: Inoculum production. In Powell CLI, Bagyaraj DJ (eds): "VA Mycorrhizas." Boca Raton: CRC Press, 1984, p 187.

146. Menge JA, Davis RM, Johnson ELV, Zentmyer GA: Mycorrhizal fungi increase growth and reduce transplant injury in avocado. Calif Agric 32:6, 1978.

147. Menge JA, Jarrell WM, Labanauskas CK, Ojala JC, Huszar EL, Johnson ELV, Sibert D: Predicting mycorrhizal dependency of Troyer citrange on *Glomus fasciculatus* in California citrus soils and nursery mixes. Soil Sci Soc Am Proc 46:762, 1982.

148. Menge JA, Johnson ELV, Minassian V: Effect of heat treatment and three pesticides upon the growth and reproduction of the mycorrhizal fungus *Glomus fasciculatus*. New Phytol 82:473, 1979.

149. Menge JA, Labanauskus CK, Johnson ELV, Platt RG: Partial substitution of mycorrhizal fungi for phosphorus fertilization in the culture of citrus. Soil Sci Am J 42:926, 1978.

150. Menge JA, Lembright H, Johnson ELV: Utilization of mycorrhizal fungi in citrus nurseries. Proc Int Soc Citriculture 1:129, 1977.

151. Menge JA, Raski DJ, Lider LA, Johnson ELV, Jones NO, Kissler JJ, Hemstreet CL:

Interations between mycorrhizal fungi, soil fumigation, and growth of grapes in California. Am J Enol Viticulture 34:117, 1983.

152. Mertz SM, Jr, Heithaus JJ III, Bush RL: Mass production of axenic spores of the endomycorrhizal fungus *Gigaspora margarita*. Trans Br Mycol Soc 72:167, 1979.

153. Miller RM, Carnes BA, Moorman TB: Factors influencing survival of vesicular-arbuscular mycorrhiza propagules during topsoil storage. J Appl Ecol 22:259, 1985.

154. Morandi S, Gianinazzi S, Gianinazzi-Pearson V: Intéret de l'endomycorhization dans la reprise et la croissance du Framboisier issu de multiplication végétative *in vitro*. Ann Amelior Plantes 29:623, 1979.

155. Mosse B: The influence of soil type and *Endogone* strain on the growth of mycorrhizal plants in phosphate deficient soils. Rev Ecol Biol Soc 9:529, 1972.

156. Mosse B: Plant growth responses to vesicular-arbuscular mycorrhizal. X. Responses of *Stylosanthes* and maize inoculation in unsterile soil. New Phytol 78:277, 1977.

157. Mosse B: Mycorrhiza in a sustainable agriculture. Biol Agric Hort 3:191, 1986.

158. Mosse B, Thompson JP: Vesicular-arbuscular endomycorrhizal inoculum production. I. Exploratory experiments with beans *(Phaseolus vulgaris)* in nutrient flow culture. Can J Bot 62:1523, 1984.

159. Mosse B, Warner A, Clark CA: Plant growth responses to vesicular-arbuscular mycorrhiza. XIII. Spread of an introduced vesicular-arbuscular endophyte in the field and residual growth effects of inoculation in the 2nd year. New Phytol 90:521, 1982.

160. Mulder D (ed): "Soil Disinfestation: Some Scientific and Practical Contributions in the Field of Soil Disinfestation." Amsterdam: Elsevier, 1979.

161. Nelsen CE, Safir GR: Increased drought tolerance of mycorrhizal onion plants caused by improved phosphorus nutrition. Planta 154:407, 1982.

162. Newbould P, Rangeley A: Effect of lime, phosphorus and mycorrhizal fungi on growth nodulation and nitrogen fixation by white clover *(Trifolium repens)* grown in U.K. hill soils. Pl Soil 76:105, 1984.

163. O'Bannon JH, Evans DW, Peaden RN: Alfalfa varietal response to seven isolates of vesicular-arbuscular mycorrhiza fungi. Can J Pl Sci 60:859, 1980.

164. Ocampo JA: Effect of crop rotations involving host and non host plants on vesicular-arbuscular mycorrhizal infection of host plants. Pl Soil 56:283, 1980.

165. Ojala JC, Jarrell WM, Menge JA, Johnson ELV: Influence of mycorrhizal fungi on mineral nutrition and yield of onion in saline soil. Agron J 75:255, 1983.

166. Ollivier B, Diem HG, Pinta M, Dommergues YR: Influence de l'infection endomycorhizienne sur la concentration en P et Zn des parties aeriennes de *Vigna unguiculata* cultive dans un sol. Colloques INRA 13:155, 1982.

167. Owusu-Bennoah E, Mosse B: Plant growth responses to vesicular-arbuscular mycorrhiza. XI. Field inoculation responses in barley, lucerne and onion. New Phytol 83:671, 1979.

168. Pacovsky RS: Micronutrient uptake and distribution in mycorrhizal or phosphorus-fertilised soybeans. Pl Soil 95:379, 1986.

169. Pairunan AK, Robson AD, Abbott LK: The effectiveness of vesicular-arbuscular mycorrhizas in increasing growth and phosphorus uptake of subterranean clover from phosphorus sources of different solubilities. New Phytol 84:327, 1980.

170. Parvathi K, Venkateswarlu K, Rao AS: Toxicity of soil-applied fungicides and gypsum to the vesicular-arbuscular mycorrhizal fungus *Glomus mosseae* in groundnut. Can J Bot 63:1673, 1984.

171. Plenchette C, Fortin JA, Furlan V: Growth responses of several plant species to mycorrhizae in a soil of moderate P fertility. Mycorrhizal dependency under field conditions. Pl Soil 70:199, 1983.

172. Plenchette C, Furlan V, Fortin JA: Growth stimulation of apple trees in unsterilised soil under field conditions with VA mycorrhiza inoculation. Can J Bot 59:2003, 1981.
173. Pope PE, Chaney WR, Rhodes JD, Woodhead SH: The mycorrhizal dependency of four hardwood tree species. Can J Bot 61:412, 1983.
174. Poss JA, Pond E, Menge JA, Jarrell WM: Effect of salinity on mycorrhizal onion and tomato in soil with and without additional phosphate. Pl Soil 88:307, 1985.
175. Powell CLl: Mycorrhizas in hill country soils. II. Effect of several mycorrhizal fungi on clover growth in sterilized soils. NZ J Agric Res 20:59, 1976.
176. Powell CLl: Mycorrhizas in hill country soils. III. Effect of inoculation on clover growth in unsterile soil. NZ J Agric Res 20:343, 1977.
177. Powell CLl: Spread of mycorrhizal fungi through soil. NZ J Agric Res 22:335, 1979.
178. Powell CLl: Inoculation of white clover and ryegrass seed with mycorrhizal fungi. New Phytol 83:81, 1979.
179. Powell CLl: Inoculation of barley with efficient mycorrhizal fungi stimulates seed yield. Pl Soil 59:487, 1981.
180. Powell CLl: Selection of efficient VA mycorrhizal fungi. Pl Soil 68:3, 1982.
181. Powell CLl: Phosphate response curves of mycorrhizal and non-mycorrhizal plants. III. Cultivar effects in *Lotus pedunculatus* Cav. and *Trifolium repens* L. NZ J Agric Res 25:217, 1982.
182. Powell CLl: Field inoculation with VA mycorrhizal fungi. In "VA Mycorrhizas." Powell CLl, Bagyaraj DJ (eds): Boca Raton: CRC Press, 1984, p. 205.
183. Powell CLl, Bagyaraj DJ: VA mycorrhizal innoculation of field crops. Proc Annu Conf Agron Soc NZ 12:85, 1982.
184. Powell CLl, Bagyaraj DJ (eds): "VA Mycorrhizas." Boca Raton: CRC Press, 1984.
185. Powell CLl, Clark GE, Verbene NJ: Growth response of four onion cultivars to several isolates of VA mycorrhizal fungi. NZ J Agric Res 25:465, 1982.
186. Powell CLl, Daniel J: Growth of white clover in undisturbed soils after inoculation with efficient mycorrhizal fungi. NZ J Agric Res 21:675, 1978.
187. Powell CLl, Groters M, Metcalfe D: Mycorrhizal inoculation of a barley crop in the field. NZ J Agric Res 23:107, 1980.
188. Rangeley A, Daft MJ, Newbould P: The inoculation of white clover with mycorrhizal fungi in unsterile hill soils. New Phytol 92:89, 1982.
189. Reeves BF, Wagner D, Moorman T, Kiel J: The role of endomycorrhizae in revegetation practices in the semi-arid west. I. A comparison of incidence of mycorrhizae in severely disturbed vs natural environments. Am J Bot 66:6, 1979.
190. Rich JR, Bird GW: Association of early-season vesicular-arbuscular mycorrhizae with increased growth and development of cotton. Phytopathology 64:1421, 1974.
191. Rives CS, Bajwa MI, Liberta AE, Miller RM: Effects of topsoil storage during surface mining on the viability of VA mycorrhiza. Soil Sci 129:253, 1980.
192. Ross JP: Effect of phosphate fertilization on yield of mycorrhizal and nonmycorrhizal soybeans. Phytopathology 61:1400, 1971.
193. Ross JP, Harper JA: Effect of endogone mycorrhiza on soybean yields. Phytopathology 60:1552, 1970.
194. Ross JP, Ruttencutter R: Population dynamics of two vesicular-arbuscular endomycorrhizal fungi and the role of hyperarasitic fungi. Phytopathology 67:490, 1977.
195. Rovira AD: Studies on soil fumigation. I. Effects on ammonium, nitrate, and phosphate in soil and on growth, nutrition and yield of wheat. Soil Biol Biochem 8:241, 1976.
196. Saif SR: The influence of soil aeration on the efficiency of vesicular-arbuscular mycorrhizas. II. Effect of soil oxygen on growth and mineral uptake in *Eupatorium*

odoratum L., *Sorghum bicolor* (L.) Moench and *Guizotia abyssinica* (L.F.) Cass. inoculated with vesicular-arbuscular mycorrhizal fungi. New Phytol 95:405, 1983.
197. Saif SR, Khan AG: The effect of vesicular-arbuscular mycorrhizal associations on growth of cereals. II. Effects on barley growth. Pl Soil 47:17, 1977.
198. Salinas JG, Sanz JI, Sieverding E: Importance of VA mycorrhizae for phosphorus supply to pasture plants in tropical oxisols. Pl Soil 84:347, 1985.
199. Salt GA: The increasing interest in 'minor pathogens'. In Schippers B, Gams W (eds): "Soil-Borne Plant Pathogens." London: Academic Press, 1979, p 289.
200. Sanders FE, Tinker PB, Black RLB, Palmerley SM: The development of endomycorrhizal root systems. I. Spread of infection and growth-promoting effects with four species of VA endophyte. New Phytol 78:257, 1977.
201. Schenck NC, Hinson K: Response of nodulating soybeans to a species of *Endogone* mycorrhiza. Agron J 65:849, 1973.
202. Schenck NC, Kellam MK: "The Influence of Vesicular-Arbuscular Mycorrhizae on Disease Development, Bulletin 798 (Technical)." Gainesville: Agricultural Experiment Stations, University of Florida, 1978.
203. Schenck NC, Smith GS: Responses of six species of vesicular-arbuscular mycorrhizal fungi and their effects on soybean at four soil temperatures. New Phytol 92:193, 1982.
204. Schubert A, Hayman DS: Plant growth responses to vesicular-arbuscular mycorrhiza. XVI. Effectiveness of different endophytes at different levels of soil phosphate. New Phytol 103:79, 1986.
205. Schwab SM, Menge JA, Leonard RT: Comparison of stages of vesicular-arbuscular mycorrhiza formation in sudan-grass grown at two levels of phosphorus nutrition. Am J Bot 70:1225, 1983.
206. Sieverding E: Influence of soil water regimes on VA mycorrhiza. III. Comparison of three mycorrhizal fungi and their influence on transpiration. Acker Pflanzenbau 153:52, 1984.
207. Siqueira JO, Hubbell DH, Kimbrough JW, Schenck NC: *Stachybotrys chartarum* antagonistic to azygspores of *Gigaspora margarita* in vitro. Soil Biol Biochem 16:679, 1984.
208. Siqueira JO, Hubbell DH, Mahmud AW: Effect of liming on spore germination, germ tube growth and root colonization by vesicular-arbuscular mycorrhizal fungi. Pl Soil 76:115, 1984.
209. Smith GS, Roncadori RW: Responses of three vesicular-arbuscular mycorrhizal fungi at four soil temperatures and their effects on cotton growth. New Phytol 104:89, 1986.
210. Smith SE, Bowen GD: Soil temperature, mycorrhizal infection and nodulation of *Medicago truncatula* and *Trifolium subterraneum*. Soil Biol Biochem 11:469, 1979.
211. Smith SE, Gianinazzi-Pearson V: Physiological interactions between symbionts in vesicular-arbuscular mycorrhizal fungi. Annu Rev Pl Physiol 38:in prep, 1988.
212. Snellgrove RC, Stribley DP: Effects of pre-inoculation with a vesicular-arbuscular mycorrhizal fungus on growth of onions transplanted to the field as multi-seeded peat modules. Pl Soil 92:387, 1986.
213. Sreeramulu KR, Bagyaraj DJ: Field responses of chilli to VA mycorrhiza on black clayey soil. Pl Soil 93:299, 1986.
214. Stahl PD, Smith WK: Effects of different geographic isolates of *Glomus* on the water relations of *Agropyron smithii*. Mycologia 76:261, 1984.
215. Strong ME, Davies FT: Influence of selected vesicular-arbuscular mycorrhizal fungi on seedling growth and phosphorus uptake of *Sophora secundiflora*. Hort Sci 17:620, 1982.

216. Swaminathan K, Verma BC: Symbiotic effect of vesicular-arbuscular mycorrhizal fungi on the phosphate nutrition of potatoes. Proc Indian Acad Sci 85:310, 1977.

217. Sylvia DM, Schenck NC: Germination of chlamydospores of three *Glomus* species as affected by soil matric potential and fungal contamination. Mycologia 75:30, 1983.

218. Sylvia DM, Schenck NC: Application of superphosphate to mycorrhizal plants stimulates sporulation of phosphorus tolerant vesicular-arbuscular mycorrhizal fungi. New Phytol 95:655, 1983.

219. Tavares M, Hayman DS: Soil pH preference of different VA mycorrhizal endophytes. In "Rothamsted Experimental Station Report for 1982, Part 1." Harpenden: Rothamsted Experimental Station, 1983, p 218.

220. Tester M, Smith FA, Smith SE: Phosphate inflow into *Trifolium subterraneum* L.: Effects of photon irradiance and mycorrhizal infection. Soil Biol Biochem 17:807, 1985.

221. Tommerup IC: Temperature relations of spore germination and hyphal growth of vesicular-arbuscular mycorrhizal fungi in soil. Trans Br Mycol Soc 81:381, 1983.

222. Vanderzaag P, Fox RL, De La Pena RS, Yost RS: P nutrition of cassava, including mycorrhizal effects on P, K, S, Zn and Ca uptake. Field Crops Res 2:253, 1979.

223. Van Nuffelen M, Schenck NC: Spore germination, penetration, and root colonization of six species of vesicular-arbuscular mycorrhizal fungi on soybean. Can J Bot 62:624, 1984.

224. Wang GA, Stribley DP, Tinker PB, Walker C: Soil pH and vesicular-arbuscular mycorrhizas. In Fitter AH, Atkinson D, Read DJ, Usher MB (eds): "Ecological Interactions in Soil: Plants, Microbes and Animals." Oxford: Blackwell, 1985, p 219.

225. Warner A: Re-establishment of indigenous vesicular-arbuscular mycorrhizal fungi after topsoil storage. Pl Soil 73:387, 1983.

226. Warner A, Gee P, Fyson A: Studies on the infectivity of fungal mycelium. In "Rothamsted Experimental Station Report for 1982, Part 1." Harpenden: Rothamsted Experimental Station, 1983, p 219.

227. Wilson JM: Competition for infection between vesicular-arbuscular mycorrhizal fungi. New Phytol 97:427, 1984.

228. Withers LA, Alderson PD (eds): "Plant Tissue Culture and Its Agricultural Applications." London: Butterworths, 1986.

229. Yost RS, Fox RL: Contribution of mycorrhizae to P nutrition of crops growing on an oxisol. Agron J 71:903, 1979.

230. Young CC, Juang TC, Guo HY: The effect of inoculation with vesicular-arbuscular mycorrhizal fungi on soybean yield and mineral phosphorus utilization in subtropical-tropical soils. Pl Soil 95:245, 1986.

Biotechnology in Agriculture, pages 175–202
© 1988 Alan R. Liss, Inc.

Electric Gene Transfer Into
Plant Protoplasts and Cells

Hiromichi Morikawa, Asako Iida, and Yasuyuki Yamada

Research Center for Cell and Tissue Culture, Faculty of Agriculture,
Kyoto University, Kyoto 606, Japan

———————◆◆———————

———————◆◆———————

I. ELECTROFUSION

A. Introduction

Electric gene transfer can be grouped into three methods, i.e., electrofusion, electroporation, and electroinjection. We will start our review with electrofusion.

Electrofusion was first reported by Senda et al. [1] in 1979 using glass microelectrodes to induce fusion of protoplasts isolated from cultured cells of *Rauwolfia serpentina*. Neumann et al. [2] independently reported successful mass electrofusion of *Dictyostelium* using parallel metal electrodes. Weber et al. [3] reported that electric fields stimulated cell fusion of yeast protoplasts by polyethylene glycol (PEG). Zimmermann and Scheurich [4] reported a high-frequency electrofusion technique by combining Pohl's dielectrophoresis with electrofusion using an electrode chamber in which interelectrode distance was 100–200 μm. In the latter method, dielectrophoretic pair formation occurred on the electrode surface and the protoplasts tended to adhere to the electrode surface on pulsation, which frequently resulted in lethal damage to the protoplasts. Later, Watts and King [5] showed that protoplasts can be paired by an AC field anywhere in a parallel electrode chamber where interelectrode distance is 5 mm, and that these pairs can be fused by subsequent application of DC pulses. Since these reports many researchers adopted this technique for electrofusion and various commercial apparatus based on this technique became available.

Electrofusion has now been used for production of somatic hybrids of higher plants (see below), yeast [6,7], moss [8], and animal cells [e.g., 9].

B. Background

Three different steps are involved in protoplast fusion: A) formation of heteroplasmic protoplast pairs to be fused, B) induction of fusion between the paired protoplasts, and C) culture of heteroplasmically fused protoplasts (heterokaryons) and selection of somatic hybrids. Step A is usually accomplished by application of an AC field (see Table I). Its effect is interpreted as dielectrophoresis [e.g., 4] or dipole-dipole attraction [5]. Application of an AC field, although useful, is not essential for electrofusion: protoplasts can effectively be paired by other means, including direct physical contact by a multiprotoplast-layer method [10,11], or by addition of chemicals to the chamber medium [12,13] with subsequent electrofusion by a DC pulse, as described below.

For step B, DC pulse(s) is applied and both exponentially decaying pulse generated by the capacitor-discharge method and square pulse have proven to be effective. The former type of pulse is characterized by the initial intensity of electric field ($E_0 = V_0/d$, where V_0 and d are the voltage applied to the electrode chamber at time zero and the interelectrode distance, respectively), and the time constant of the decay (τ), which is equal to CR where C is the capacitance of the capacitor and R the resistance of the

TABLE I. Experimental Conditions for Production of Somatic Hybrids by Electrofusion in Plants

Plant material	Pair formation	DC pulse condition				Medium	Selection method	Hybrid	References
		kV/cm	Δt^a (μs)	CR^b (μs)	n^c				
N. tabacum (mesophyll) + N. plumbaginifolia (suspension culture)	AC, 600 kHz 15 V/chamber	50 V/chamber	50	—	2	No added salts	Pick up	Plantlets	24
N. tabacum (suspension culture) NR-deficient cell lines	AC, 900 kHz 400 V/cm	1.2–1.5	50	—	1	No added salts	Minimum medium	Calli Shoots	30
N. glauca(mesophyll) + N. langsdorffii (mesophyll)	Multiprotoplast layer	1.0	—	100–200	1	+ 2.5 mM $CaCl_2$	Hormone-free medium	Fertile plants	10,32
N. plumbaginifolia (mesophyll and suspension culture) NR-deficient cell lines	AC, 1 MHz 150 V/cm	1.2–2.0	—	50	1	No added salts	Pick up	Calli	31
N. glauca + N. langsdorffii (mesophyll)	1–10% PEG	1.0	100	—	1	+ 1 mM $MgCl_2$, 50 mM NaCl, 0.75 mM spermine, 0.01% Tween 80 and 1–10% PEG	Hormone-free medium	Calli	13
S. tuberosum + S. phureja (shoots)	AC, 1 MHz 150 V/cm	1.2–2.0	50	—	1	No added salts	Pick up	Plants	33
S. tuberosum (suspension culture) amino acid analog-resistant cell lines	AC, 1.5 MHz >233 V/cm	>3.3	10–1,000	—	1	+ 0.15–1.0 mM $CaCl_2$	Medium containing amino acid analogs	Calli	35

[a]Duration of square pulse.
[b]Time constant of exponentially decaying pulse.
[c]No. of pulses.

chamber. The intensity of electric field at time t (E_t) is given by the following equation:

$$E_t = E_0\exp(-t/\tau) \tag{1}$$

and

$$E_{t=\tau} = 1/eE_0 \fallingdotseq 0.37E_0$$

The square pulse is characterized by the field strength and duration (Δt). It is possible that there is a threshold value for both DC field intensity and the pulse width (duration or time constant) for the induction of protoplast fusion, but this has not been demonstrated conclusively.

The electrical conductivity of the chamber medium modulates the effects of the DC pulse. Thus, the concentration and composition of the ions, or electrolytes, in the chamber medium are important factors. These are also important for the stability of protoplast membranes, as well as the intracellular biochemistry of protoplasts. Usually, in addition to an appropriate sugar or sugar alcohol which acts as an osmoticum, divalent cations such as Ca^{2+} or Mg^{2+} are added to the medium. The effects of these cations on fusion efficiency, however, are still controversial (see below).

Selection of hybrids (step C) can be made by the mass culture of all fusion products under appropriate selection pressure, based on the autotrophy of the hybrids or the auxotrophy of parental protoplasts. Additionally, heterokaryons can be isolated or enriched prior to culture by a manual selection (or picking up) with a glass micropipette, or by an automatic flow-sorting system. Manual selection of electrically produced heterokaryons is tedious, but several somatic hybrids were established by this technique, followed by microculture of selected heterokaryons (see Table I). There is no universal method for selection or purification of heterokaryons from the fusion products at present. In this respect the development of efficient flow-cytometric system for purification of heterokaryons will be most promising.

The voltage across the membrane (V_m) of protoplasts exposed to an external electric field (E) is calculated by the following equation [14,15]:

$$V_m = 1.5rE\cos\theta \tag{2}$$

where r is the radius of the protoplast and θ the angle between the normal to the membrane surface and the field direction; i.e., the voltage across the membrane is a maximum in the direction of the electric field and $1.5rE$. Thus, when an electric field of 1 kV/cm is applied to the protoplast of $r = 40$ μm, the maximum voltage across the membrane is calculated to be 6 V. This value is 30–60 times higher than the resting membrane potential of plant cells. The critical voltage for membrane breakdown (for pore formation or membrane fusion) in plants is reported [15] to be the order of 1 V per membrane.

There is no doubt that the cell membranes are "activated" by electric field pulses so that cell membranes become easy to fuse each other or they are permeabilized to allow the entry of macromolecules which normally are not permeable to them. However, it is not self-evident that this "activated state" of membranes is equivalent to the so called "electroporated" state. For example, Mehrle et al. [16], Sowers [17], and Sowers and Lieber [18] reported that the electropermeabilization of cell membranes has a polarity, and this suggests that electropore formation is unlikely to be involved in membrane electrofusion. Also, virus particles as large as $180 \times 3,000$ Å are reported [19–21] to be introduced into protoplasts by electroporation. This suggests that the electric field activates endocytotic activity of the plasma membrane of protoplasts rather than simple uptake through electropores.

If the voltage necessary to activate the cell membrane were the same among various plant species, then Equation 2 suggests that the electric field strength of the fusion pulse should be varied as a function of the size of the protoplasts. It is reported [22] that mesophyll protoplasts are more readily fused than protoplasts of suspension cultured cells. This indicates that other factors in addition to size, such as the character of the membrane itself, modulates the effectiveness of the applied electric field.

The most distinct advantage of electrofusion, as compared with chemical fusion methods, would be the ease and reproducibility of the experimental procedures. Other advantages of electrofusion are discussed in recent reviews [e.g., 23–25]. Possible effects of a high DC fusion pulse on the nuclear and organellar DNA, however, need to be studied in future. Although no detailed studies have been reported on possible differences in the hybrid isolation yield between electrofusion and chemical fusion, electrofusion is reported to be free of cytotoxicity, which often is associated with chemical fusogens such as PEG or polyvinyl alcohol (PVA). Tur-Kaspa et al. [26] reported the introduction of foreign genes into primary culture of animal cells by electroporation. Transfection of the cells by calcium phosphate or diethylaminoethyl (DEAE)-dextran was unsuccessful because of the cytotoxicity of the chemicals. Successful transfer of disease resistance into potato [27] and eggplant [28] by somatic hybridization by chemical fusion has been reported. This result is also encouraging for the study of gene transfer by the electrical method.

C. Advances in Studies on Somatic Hybridization by Electrofusion in Plants

Table I summarizes recent studies on the production of somatic hybrids of higher plants by electrofusion. All the successfully hybridized plant materials belong to Solaneceae involving five *Nicotiana* species and two *Solanum* species.

Bates and co-workers [29] first showed that electrofusion treatment itself is not lethal by successful colony formation from electrically treated protoplasts. Subsequently they produced [24] somatic hybrid plantlets after electrofusion between *Nicotiana plumbaginifolia* and *N. tabacum* protoplasts. Hybridity was confirmed by the analysis of the esterase isozyme pattern. Kohn et al. [30] also succeeded in the production of somatic hybrids from electrically fused protoplasts from two complementing nitrate reductase (NR)-deficient cell lines of *N. tabacum* and of *N. plumbaginifolia*. The somatic hybrids were selected on a medium containing nitrate as a sole nitrogen source. The selected hybrid calli had a NR activity and some regenerated shoots [30]. Puite et al. [31] produced somatic hybrid calli from electrically fused protoplasts that were isolated from two complementary NR-deficient mutants of *N. plumbaginifolia*.

Complete, fertile somatic hybrid plants were first electrically produced between *N. glauca* and *N. langsdorffii* by Morikawa et al. [10,32]. From 45,000 mesophyll protoplasts of each species, they obtained more than 96 green hybrid calli that grew vigorously on a hormone-free medium after 10 wk of culture. The hybrid isolation yield was about 0.2%. All hybrid calli regenerated shoots after about 4 mo of culture, and some formed roots. After more than 8 mo of culture some regenerants had developed to mature plants, flowered, and set seeds. Seven such plants were obtained. Hybridity of these plants was confirmed by electrofocusing analysis of fraction I proteins isolated from the leaves of these plants (see also below). Chapel et al. [13] also have obtained somatic hybrids between *N. glauca* and *N. langsdorffii* by protoplast electrofusion. Hybridity was confirmed by analysis of fraction I proteins. The hybrid isolation yield was reported to be 0.14%.

Puite et al. [33] first succeeded in production of somatic hybrid *Solanum* plants by electrofusion; they electrofused protoplasts from shoots of *Solanum tuberosum* and *S. phureja* and manually selecting the heterokaryons (visual identification of the heterokaryons was facilitated by fluorescein diacetate staining of the protoplasts from one of the parents, which was grown on a herbicide to induce bleaching of the chlorophyll). They obtained 18 putative hybrid plants that were grown to maturity. The hybridity of these plants was established by determining nuclear DNA content and by karyotype analysis [33]. Fertility of the hybrids is not addressed. Recently, Pijinacker et al. [34] reported the elimination of specific chromosomes in these *Solanum* hybrids. De Vries et al. [35] carried out intraspecific hybridization between amino acid analog-resistant cell lines of potato (*S. tuberosum*) by electrofusion. Hybrid calli were selected in the presence of the amino acid analogs (S-aminoethylcysteine and 5-methyltryptophan).

In order to induce pair formation or aggregation of protoplasts prior to application of fusion (DC) pulse, most researchers have employed AC-field-induced movement (dielectrophoresis or dipole-dipole attraction) of protoplasts as developed by Zimmermann and Scheurich [4], Richter et al. [36], and Watts and King [5] (see Table I). Morikawa et al. [10,32] and Yamada et al. [11] did not use an AC field for protoplast pair formation but rather a "multiprotoplast-layer method" (where up to 50 layers of protoplasts were formed in the electrode chamber to give rise to membrane-membrane contact). Chapel et al. [12,13] also did not use an AC field, but they chemically aggregated protoplasts by adding spermine or PEG to the chamber medium. These results clearly indicate that an AC field is not essential for production of hybrid plants by electrofusion.

Both capacitor-discharge and square pulse have effectively induced fusion. A rather narrow range of electrical conditions has been used for somatic hybridization (see Table I). The field strength varies from 1.0 to 3.3 kV/cm, and the duration or time constant is 50–200 µs, and mostly single pulse is applied. In contrast, the chamber media employed are quite different. These vary from a medium with no added salts (where electric resistance would be several tens of kΩ) to a medium containing more than 50 mM of electrolytes (electric resistance of this medium would be less than one-hundredth that of salt-free medium). This means that the effective value for the field strength would be different among the conditions listed in Table I. Optimization of the DC fusion pulse in terms of the field strength, width of the pulse, number and interval of the pulse, shape (decaying or square) of the pulse, concentration and composition (particularly of divalent cations, see below) of the chamber medium, temperature and so forth, has yet to be done systematically.

Addition of 0.1–2.5 mM Ca^{2+} (as $CaCl_2$) [5,10,37] or 0.75–1.0 mM spermine [12,37] to the chamber medium is reported to increase the viability of fused protoplasts and the fusion frequency. On the other hand, Chapel et al. [12] reported that Mg^{2+} stimulates electrofusion but Ca^{2+} is inhibitory (see also below discussion on the effects on electroporation).

Note that three of seven hybrids listed in Table I were produced by manual selection of heterokaryons from fusion products and by subsequent micro-culture. Manual selection is quite useful for the isolation of heterokaryons [25]. However, it is very difficult to obtain hybrid colonies when the plating efficiency of parental protoplasts is low [11].

Technical improvements for electrofusion in instruments have been undertaken by various researchers [e.g., 23,38,39]. One of the most important steps for electrofusion is the formation of protoplast pairs to be fused. No method is available for specific heteroplasmic protoplast pair formation, except manual pair formation using microelectrodes connected to micromanipulators [40,41] or a more sophisticated hydraulic device [42]. However, the numbers of protoplasts that can be handled with these methods are quite limited, and thus procedures leading to preferential heterokaryon production also result in drastic cuts in the yields of fusion products [38]. Tempelaar and Jones [22] found that mesophyll protoplasts fuse more readily than suspension-culture protoplasts and showed that the yield of binary fusion products can be increased by changing the ratio of mesophyll to suspension protoplasts [43]. Hampp and Steingraber [44] reported electro-fusion between vacuolate and evacuolate mesophyll protoplasts. The differ-ence in buoyant density among the vacuolate, evacuolate, and heteroplasmi-cally fused protoplasts was utilized for separation of the heterokaryons by density gradient centrifugation [45].

Recently, we studied the effects of physical parameters including the osmotic potential and density of the chamber medium and the hydrophilic/hydrophobic coating of the bottom surface of the chamber, upon protoplast electrofusion. We have found that there is an optimum osmotic potential and density (or buoyant force) of the medium for electrofusion and also that the hydrophilic coating of the bottom surface with Gellan gum or polyacrylamide gels stimulated electrofusion, while hydrophobic siliconization was entirely inhibitory [45a].

It is known that pretreatment of animal cells with proteolytic enzymes promotes their electrofusion [46,47]. Pretreatment of plant protoplasts with various proteolytic enzymes such as protease and pronase also is reported to increase fusion frequency (although this treatment exerted damaging effects on the protoplasts) [48]. We have found that in interkingdom electrofusion between barley mesophyll protoplasts and cultured mouse lymphoblast cells L5178Y, pretreatment of the mouse cells with 1.3% Cellulase Onozuka R10 and 1.3% Macerozyme R10 for 90 min at 36°C was effective (Morikawa, Okada, and Senda, unpublished results). The positive effect of these cell-wall-degrading enzyme preparations is somewhat surprising, but it is not clear that the observed effects are due to the glycosidase activities and not to proteolytic activities of these preparations (proteolytic activity has been demonstrated in these crude enzyme preparations). These results indicate that modification of the cell membrane surface affects the electrofusion process. Recently, calmodulin antagonists and cytoskeleton inhibitors also were shown to affect electrofusion process, indicating a possible involvement of calmodulin and cytoskeleton in protoplast electrofusion [49].

In electrofusion of protoplasts from cultured cells of *Coptis japonica* and *Euphorbia millii* using platinum microelectrodes, we showed [41] that the process involved two stages, cellular and vacuolar fusion, which are characterized, respectively, by transient wrinkling of the protoplast membrane and the formation of dark red precipitates. The transient wrinkling of the membrane suggested that the protoplast fusion occurred under constant volume, which resulted in a transient formation of an "extra" surface area during protoplast electrofusion. The vacuoles of *Euphorbia millii* and *Coptis japonica* contain large amounts of anthocyanins and berberine, respectively [11,25,41]. The mechanism of the precipitate formation upon vacuolar fusion was considered as follows: the cell sap of the latter may have a higher pH than that of the former, and anthocyanins precipitated by the increased pH in the fused vacuoles.

D. Production of Fertile Somatic Hybrid Plants by Electrofusion

Figure 1 shows hybrid plant formation in electrofused protoplasts of *Nicotiana glauca* and *N. langsdorffii*. We have been culturing more than 90 hybrid clones on hormone-free medium that were obtained by electrofusion in October 1984, as reported previously [10]. After more than 8 mo of culture, some hybrid clones formed small plantlets, as shown in Figure 1B, which eventually developed to mature, fertile plants (see Fig. 1C,D). The morphology of young (C) and mature (D) plants was distinct from parental plants but similar to that of amphidiploid plants obtained from colchicine treatment of sexual hybrids of *N. glauca* and *N. langsdorffii* [32]. At present we have obtained more than seven complete hybrid plants like the one shown in Figure 1D, and all have set seeds upon self pollination.

Table II summarizes cytological data obtained with seven somatic hybrid plants. The electrofocusing analysis of the subunits of the fraction I proteins extracted from the leaves of the plants revealed that all plants had small subunit (SS) polypeptides of both the *N. glauca* type (G) and *N. langsdorffii* type (L). This result indicates that the information encoded on the two different nuclear genomes of parental *Nicotiana* species was expressed in all of these hybrid plants and thus confirm their hybridity.

The chromosome number (2n) of the plants was between 60 and 66. The parental diploid chromosome numbers are 24 for *N. glauca* (GG) and 18 for *N. langsdorffii* (LL). Simple addition would give a chromosome number of

Fig. 1. Hybrid plant formation in electrofused cells of *Nicotiana glauca* and *N. langsdorffii*. A: Homo- or heteroplasmically fused protoplasts about 1 h after electrofusion. B: Plantlets regenerated more than 8 mo after electrofusion. C: Young somatic hybrid plant 18 mo after electrofusion. D: Mature somatic hybrid plant 19 mo after electrofusion.

42, while the products of triple fusion would have a chromosome number of 60 or 66:LL + GG = 42 chromosomes; LL + LL + GG = 60 chromosomes; LL + GG + GG = 66 chromosomes. The present results strongly suggest that the hybrid plants are the products of triple fusion and that the loss or increase of chromosomes during culture gave rise to aneuploid types.

In Table II, each of the clones 17, 54, and 56 are derived from a single heterokaryon. The subclones 17-1, 17-2, 56-1, and 56-2 were separately subcultured during callus (teratoma) culture for more than 8 mo before regeneration of plantlets and the subclones 56-1 and 56-2, in each series, regenerated two plants from the same culture bottle (56-1, No.1; 56-1, No.2;

TABLE II. Cytological Information on Seven Mature Hybrid Plants Produced by Electrofusion of Protoplasts of *N. glauca* and *N. langsdorffii*

Clone No.	2n	Subunits of FIP[a]		Viable seeds produced
		SS	LS	
17-1	66	G + L	L	+
17-2	63	G + L	L	+
54	66	G + L	L	+
56-1, No. 1	66	G + L	L	+
56-1, No. 2	64	G + L	L	+
56-2, No. 1	60	G + L	L	+
56-2, No. 2	66	G + L	L	+

[a]FIP, SS, LS, G and L = fraction I protein, its small and large subunit, and *N. glauca* and *N. langsdorffii* types, respectively. Reproduced from Morikawa et al. [32] with permission.

56-2, No.1; and 56-2, No.2). Thus, the presence of chromosomal aberration among subclones of the clone 17 or 56 indicates that chromosome loss or increase readily occurs during the culture (and possibly during plantlet regeneration) of the hybrid calli.

The present result indicates that the large subunit polypeptides of the fraction I proteins (LS) was only *N. langsdorffii* type in all hybrid plants (see Table II). Since the gene for this subunit polypeptides is on the chloroplast genome, the *N. langsdorffii* chloroplasts may have been dominant in number over *N. glauca* chloroplasts in the heterokaryon; i.e., the putative triple fusion may have been "LL + LL + GG" type, and this dominance of the *N. langsdorffii* chloroplasts was retained in the hybrids. It is also possible that the triple fusion was "LL + GG + GG" type, and the *N. glauca* chloroplasts were eliminated from the hybrids during culture by unknown mechanism. Pertinent studies including karyotype analysis are currently being made.

Our interspecific hybrid plants in *Nicotiana* appeared to be the products of a triple fusion as described above. Previous authors also reported multiple fusion products in electrofusion of *Solanum* protoplasts after AC field induced pair formation [33] and in PEG-induced fusion in *Nicotiana* [see references in ref. 32] and *Brassicca* [50] protoplasts. Also Bates [38] reported that more than 40% of the heterokaryons electrically produced after AC-field-induced pair formation had more than three nuclei. These results suggest that a) the multiprotoplast-layer method is not the cause for the multiple-fusion phenomenon and that b) multiple-fusion products may have a high regenerative ability. This also will be an interesting aspect for future study in cell fusion.

II. ELECTROPORATION AND ELECTROINJECTION

A. Introduction

In 1968 Sale and Hamilton [14] reported that, when high DC electric field pulses were applied to bacterial protoplasts, spheroplasts, or erythrocytes, their membranes lose the property of semipermeability. Auer et al. [51] reported that radioactive DNA or RNA can be introduced into human red cells by using electric field pulses (dielectric breakdown). Neumann et al. [52] succeeded in producing stable transformants by electric gene transfer: they added plasmids containing the herpes simplex thymidine kinase (TK) gene to a suspension of mouse L cells deficient in the TK gene, and the cells were then exposed to electric field pulses, followed by culture in the selection medium containing hypoxanthine, aminopterin, and thymidine (HAT medium). The transformation efficiency was 9.5×10^{-5}. They proposed an electroporation model to explain this cross membrane transport of the macromolecules in response to the electric field.

In the field of plant genetic engineering, most transformation studies have employed *Agrobacterium tumefaciens,* in which part of tumor-inducing (Ti) plasmid, called T-DNA, is transferred to the chromosomal DNA of (dicot) host plants or tissues, to introduce foreign DNA into the nuclear genome. Davey et al. [53] and Krens et al. [54] first transformed plant protoplasts with isolated Ti plasmid DNA (''naked'' DNA) using chemicals (poly-L-ornithine and PEG, respectively). Paszkowski et al. [55] showed that chimeric plasmid DNA, free of Ti-plasmid DNA fragments, can be introduced into the cells and integrated into the chromosomal DNA of host plants by means of a protoplast transformation technique of Krens et al. [54]. They called this technique ''direct gene transfer.'' This result shows that plant cells can be transformed without the help of *Agrobacterium* or the Ti-plasmid. Since this report, there have been several reports of the stable transformation of plants by the direct gene transfer [56–63]. The transformation efficiency so far reported by this method is the order of 10^{-4}–10^{-3}, which is similar to or higher than that reported by electroporation (see below).

Fromm et al. [64,65] first showed that direct gene transfer also can be carried out by an electric field pulse, or electroporation. Since then a number of reports have been published on electroporation of various genes, including chimeric plasmid DNA and RNA into plant protoplasts, as described below.

Hashimoto et al. [66] and Morikawa et al. [67] first succeeded in introduction of, respectively, plasmid DNA and viral RNA into intact cells of yeast and free mesophyll cells of tobacco using electric field pulses. They coined the term *electroinjection* for this technique of introduction of genetic materials into cells through cell walls and membranes. Electroinjection avoids the difficulties inherent in protoplast transformation, and in principle (as a physical method), it should be applicable to all plant species. Details are described below.

B. Advances in Studies on Electroporation and Electroinjection

Table III summarizes recent studies on electroporation and electroinjection. Plasmid DNA, intact and engineered viral RNA, and virus particles have been successfully introduced into, and expressed transiently or stably in, mono- and dicot plant protoplasts by electroporation. Plasmid DNA and viral RNA are successfully transferred, respectively, into intact yeast cells [66] and free mesophyll cells [21,67] by electroinjection, but transfer of chimeric plasmids into free cells by this method was reported [68] to be unsuccessful.

The chimeric plasmids that used for transformation of higher plants contain either bacterial Tn5 neomycin phosphotransferase type II (NPTII) or bacterial Tn9 chloramphenicol acetyltransferase (CAT) gene as a selectable marker under the control of promoters and polyadenylation signals of various origin. The enzyme encoded in NPTII gene inactivates aminoglycoside antibiotics (neomycin, kanamycin, or G-418) by phosphorylation and confers stable resistance to them in bacteria, mammalian cells, and plant cells [e.g., ref. 55 and references therein). Transient expression of the CAT gene has been very useful in the study of promoter function in mammalian cells and plant cells [see references in ref. 64]. The promoters most widely utilized in the plasmids are of Ti plasmid or virus origin:nopaline synthesis (NOS) and mannopine synthesis (MAS) gene promoter and 1' promoter of 1'–2' dual promoter (T1') of Ti plasmid and 35S and 19S promoter of cauliflower mosaic virus DNA. Promoters of plant and animal origin also have been utilized successfully. These include promoters from the carrot phenylalanine ammonialyase gene (PAL), the maize alcohol dehydrogenase 1 gene (ADH), and the maize zein gene (ZEIN) and the *copia* long terminal promoter of *Drosophila* (COPIA).

The polyadenylation signals utilized in the chimeric genes were derived from the following genes: the NOS gene, octopine synthesis gene (OCS), and gene 7 (g7) of Ti plasmid, the gene of the small subunit of ribulose 1,5-bisphosphate carboxylase (rbcS), the gene from simian virus SV40 (SV40), and the gene from 19S polyadenylation signals of cauliflower mosaic virus DNA (19S).

For transformation of yeast cells or protoplasts, plasmid YEp13 or YRp7 carrying *LEU2*$^+$ or *TRP1*$^+$ gene has been used.

So far, most of the studies of electroporation and electroinjection have

TABLE III. Experimental Conditions for Electroporation and Electroinjection in Plants

Genetic material			Plant material	Electric condition				Medium		References
Construction or name	Form	Concentration (μg/ml)		kV/cm	Δt^a	CR^b	n^c	mM^d	R^e	
Electroporation (into protoplasts)										
Plasmid DNA										
NOS-CAT-NOS	sc	10	Carrot, maize	0.875	—	—	1	165	—	64
35S-CAT-NOS	sc	10	Tobacco							
Ti	—	10 + carrier	Carrot	0.5–3.8	6 μs	—	6	271	—	73
19S-NPTII-19S	c & 1	0.1–100 + carrier	Tobacco	1.25–1.5	—	10 μs	3	6 (+ >13% PEG)	1 kΩ	74
TRP1$^+$ gene	—	10	Yeast	10	—	50 μs	1	20	50 Ω	86
35S-NPTII-NOS	sc	50	Maize	0.5	—	2–4 ms	1	146	—	65
NOS-CAT-NOS	sc & 1	10	Tobacco	0.75	—	6 ms	1	75	60 Ω	72
35S-CAT-NOS	sc & 1									
NOS-NPTII-										
NOS-NPTII-NOS	1	20	Tobacco	2.0	—	250 μs	1	0	25 kΩ	76
NOS-CAT-NOS	sc	1–500 + carrier	Carrot	0.875	—	10 ms	1	164	—	69
35S-CAT-NOS										
PAL-CAT-NOS										
35S-CAT-rbcS	—	20 + carrier	Wheat, rice Sorghum	2.5	0.1 ms	—	2	150	—	70
Copia-CAT-SV40	—									
35S-CAT-NOS	—	10–200	Maize	0.625	—	—	1	160	—	71
NOS-CAT-NOS										
ADH-CAT-NOS										

Construct	Form	Amount	Species							Ref.
NOS-NPTII-NOS 35S-NPTII- 35S-CAT- 19S-CAT-19S	—	20	Mono- and dicots	—	—	—	—	165	—	87
NOS-NPTII-NOS	sc	10	Soybean	0.375	—	>45 ms	1	>75	—	77
19S-NPTII-19S	c & l	10 + carrier	Tobacco N. plumbaginifolia	1–2.2	—	10 μs	3	>6 1 kΩ (+ 8–22.5% PEG)	—	75
MAS-CAT-SV40 ZEIN-CAT-SV40	sc & l	10–50	Carrot	1.6–2	—	—	1	>100 (+ 2.5–13% PEG)	—	78
Tl'-CAT-g7	—	50 + carrier	Moth bean	0.5–2	10 μs	—	3	0	—	68
RNA										
TMV CMV	—	10	Tobacco	5–10	90 μs	—	1–9	0	—	82
TMV CMV	—	<50	Tobacco Vinca rosea	0.75	—	6 ms	1	75	60 Ω	19
TMV	—	10	Tobacco	0.55–0.8	50 μs	—	1–10	0	—	39
BMV CCMV	—	5–30	Tobacco N. plumbaginifolia	2.5	5–10 μs	—	1	0	—	20
TMV Transcript of TMV cDNA	—	0.06–2.5	Tobacco	0.75	—	—	1	75	—	81

(continued)

TABLE III. Experimental Conditions for Electroporation and Electroinjection in Plants (Continued)

Genetic material			Plant material	Electric condition				Medium		References
Construction or name	Form	Concentration (μg/ml)		kV/cm	Δt^a	CR^b	n^c	mM^d	R^e	
Virus particle										
TMV CMV	—	500	Tobacco	0.75	—	6 ms	1	75	60 Ω	19
BMV CCMV	—	50	Tobacco N. plumbaginifolia	2.5–5	—	5–10 μs	1	0	—	20
TMV	—	50–500	Tobacco	0.67	10 ms	—	1	0	—	21
Electroinjection (into intact or free cells)										
LEU2⁺ gene	sc	1.7	Yeast	5.0	—	10 ms	3	0 (+ 35% PEG)	10 kΩ	66
TMVRNA	—	10	Tobacco	0.65	—	4 μs	1	2.5	2 kΩ	67
TMVRNA	—	—	Tobacco	0.67	0.1–0.5 ms	—	1	0	—	21
NOS-NPTII-OCS	—	33 + carrier	Moth bean	1–8	10 μs	—	3	>63 (+ 8% PEG)	—	68
T1′-CAT-g7	—	50 + carrier	Moth bean	0.5–8	10 μs	—	3	0	—	68

[a]Duration of square pulse.
[b]Time constant of exponentially decaying pulse.
[c]No. of pulses.
[d]Total concentration of the electrolytes in the chamber medium.
[e]Electric resistance of the medium.

been on transient transformation or transfection. Interestingly, transient expression of the CAT gene electroporated into carrot protoplasts was inhibited by cotransformation of plasmids containing the antisense CAT gene, indicating that transcription of antisense RNA effectively blocks the expression of target genes in plants as observed in bacteria and animal systems [69].

Ou-Lee et al. [70] reported that the same level of expression occurred in monocot protoplasts when the CAT gene was fused to either the 35S promoter or *Drosophila* promoter (COPIA). Howard et al. [71] reported that the expression of the CAT gene linked to the ADH promoter (known to be endogenously induced by anaerobic stress) is anaerobically regulated. Okada et al. [72] reported that the transient expression of the CAT gene or stable expression of the NPTII gene was 3–4 times or 2–8 times, respectively, higher than when it was introduced into mitotic protoplasts, as compared to other phases of the cell cycle.

Among the studies listed in Table III, there are seven reports [65,72–77] of stable transformation by electroporation. Langridge et al. [73] showed that the Ti plasmid can be electroporated into carrot protoplasts and that transformed plantlets can be regenerated from them which continue to express Ti-encoded genes. Fromm et al. [65] first succeeded in obtaining stable transgenic calli by electroporation using a chimeric 35S-NPTII-NOS gene. Riggs and Bates [76] and Negrutiu et al. [75] produced fertile transgenic tobacco plants by electroporation, with a transformation efficiency of $2.2-7.7 \times 10^{-4}$, which is very close to the efficiency obtained by the chemical method for direct gene transfer using PVA [56]. In the case of the Negrutiu et al. [75], however, 8–22.5% PEG were included in the electroporation chamber medium, and thus the transformation in this case is not simply attributable to the effect of the electric field (see discussion below). Christou et al. [77] obtained soybean roots stably transformed with kanamycin resistance gene by electroporation.

Supercoiled (sc), circular (supercoiled or open circular, c), and linear (1) forms of the plasmid DNA have been used for transient and stable transformation, as described in Table III. The effects of the form of the plasmid on the transformation efficiency are still controversial. For example, linear DNA has been reported [74,75] to be 2.5–10-fold more efficient than circular DNA in transformation of tobacco with NPTII gene. On the other hand, supercoiled DNA has been reported to be more efficient than linear DNA in the transformation of carrot with NPTII gene [78], while the two forms of DNA showed no difference in the efficiency of transformation of tobacco with the CAT gene [72]. Kanamycin-resistant transgenic tobacco plants [76] or calli [72] were obtained with linear DNA and kanamycin-resistant transgenic maize calli [65] and soybean roots [77] with supercoiled DNA.

For the entry of DNA molecules into the cells through cell walls and membranes, the most compact supercoiled DNA could be expected to be more effective than the other two conformations of DNA. However, for the expression of the introduced genes or integration of exogenous DNA, other forms of DNA may be more efficient. Recently, Werr and Lörz [61] reported that in transformation of wheat protoplasts with a supercoiled chimeric plasmid DNA carrying NPTII gene, the conformation of the introduced plasmid DNA changed from supercoiled towards the open circular and linear forms in the cells; only the latter forms were found 10 days after transformation. Transient expression of introduced NPTII gene is believed [61] to occur from the extrachromosomal plasmid copies. In animal cells it is reported that linear molecules transform more efficiently than circular molecules, but the detailed integration mechanisms of exogenous DNA into host genomic DNA is not known [e.g., 79].

The concentration of plasmid DNA used for electroporation is usually 10–50 μg/ml, with high concentrations (more than 100 μg/ml) of plasmid DNA reported to be toxic. In addition, carrier DNA such as calf thymus DNA, also is added (50–100 μg/ml) to the medium [68–70, 73–75], although some researchers report [78] that better transformation occurs in the absence of carrier DNA. Possible effects of carrier DNA upon transformation needed to be studied. Carrier DNA was integrated into host chromosomal DNA in one case [80] but not in an other [73].

Viral RNA or virus particles also have been introduced successfully into and expressed in plant protoplasts or free cells as determined by staining with virus-specific fluorescent antibody. Recently, Watanabe et al. [81] reported that very low levels (as low as 0.06–2.5 μg/ml) of tobacco mosaic virus (TMV) RNA or in vitro transcript of TMV cDNA can be introduced and expressed in tobacco protoplasts. Okada et al. [19] showed that 50% of the protoplasts were infected, even when TMV RNA was added 10 min after pulsation. However, Watts et al. [20] reported that no infection was detected if TMV RNA was added as little as 10 s after pulsation. They also reported [5,20] that positively charged brome mosaic virus (BMV) infected readily but negatively charged cowpea chlorotic mottle virus (CCMV) only poorly.

As in the case of electrofusion, both the exponentially decaying and square pulses have proven to be effective in electroporation and electroinjection. However, in contrast to electrofusion, the reported optimum electrical conditions for electroporation vary greatly; the field intensity employed for higher plants varies from 0.4 to 2.5 kV/cm, the duration or time constant of the pulse from 6 μs to 50 ms, and the number of the pulses from 1 to 10. A detailed study has yet to be made on the possible contribution of protoplast fusion to the uptake of genetic materials into cells by electroporation. Based on our comparative experiments of electroporation and electroinjection using protoplasts and protoplast-derived cells, we tend to think that the electrical conditions optimal for electroporation or electroinjection are slightly weaker than those for electrofusion (in preparation).

The media employed during the electrical pulse also vary greatly. These range from a salt-free medium to a medium containing 270 mM of electrolytes (where the electric resistance of the medium varys from 25 kΩ to 50Ω). The reasons for this discrepancy are not apparent. Systematic, fundamental studies on the electrical and nonelectrical (medium) parameters which affect the efficiency of electric gene transfer are needed before the experimental conditions can be optimized.

The presence of PEG in the medium is sometimes required [66,78] or even essential [74,75] for electroporation and electroinjection. It has yet to be investigated systematically whether there is any synergistic interaction between electric field and PEG treatment. It has been reported [68,75] that an electrical pulse gave no additional transformants when protoplasts were transformed with PEG and $MgCl_2$. Also, Watts et al. [20] reported that electroporation of viral RNA was at least as effective as inoculation in the presence of PEG. However, in the case of electroinjection of plasmid DNA into intact yeast cells, treatment with PEG in the absence of an electric field was not enough to transform the cells (see below).

The presence of $MgCl_2$ in the chamber medium is considered essential for electroporation by various authors [73–75], and Ca^{2+} is toxic [20] or has no effect [82], although the latter cation is known to be important in stabilizing the cell membrane structure. It is reported [83] that injection of Ca^{2+} into *Xenopus* eggs that are arrested at S phase by an exogenous inhibitor results in the resumption of replication of foreign plasmids as well as the endogenous nuclear DNA. Thus, the effects of divalent cations in the medium should be evaluated at least from two points, i.e., their effects on cell membranes and on intracellular biochemistry.

Medium without added salts is recommended for the chamber medium by a number of researchers because the salts in the medium are co-introduced into the cells by the electric field, and this results in a change in intracellular ionic balance [20,21,39,66,68,76,82].

Most researchers electroporate at low temperature (ca. 0° C). The rationale for this is that low temperature may a) suppress the activity of possible nucleases in the medium released from the protoplasts or cells, b) protect the cells from possible damage resulting from Joule heating, and

c) the activated ("electroporated") state of membrane may last longer at a lower temperature [52,84].

C. Electroinjection of Plasmid DNA Into Intact Yeast Cells

The first demonstration of electroinjection was reported [66] on introduction of plasmid YEp13 containing the *LEU2*$^+$ gene, which encodes the leucine biosynthetic enzyme β-isopropylmalate dehydrogenase, into intact cells of a yeast strain (*Saccharomyces cerevisiae* KK4), which is a double mutant in the *LEU2* gene. Logarithmic stage yeast cells were washed with sterile distilled water and suspended in water. A 50-μl portion of this cell suspension (2.5×10^9 cells/ml) was mixed with 10 μl of plasmid DNA solution (200 μg/ml) and 60 μl of 70% PEG 4000 in a 1.5-ml Eppendorf tube. After standing for 1 hr at room temperature, about 40 μl of the cell suspension was placed in a parallel-electrode chamber [10]. The electric field pulse was applied using a capacitor-discharge method [25,85, see also above], as a single pulse, or in trains of successive pulses at intervals of 5 s. The initial intensity of the pulses was changed from 2.5 kV/cm to 10 kV/cm, and the capacitance from 0.1 μF to 10 μF. The resistance of the cell-DNA mixture in the chamber was about 10 kΩ. After application of the pulses, the cell-DNA mixture was allowed to stand for 1 hr. The cells were returned to an Eppendorf tube and 1 ml of distilled water was added to this tube and centrifuged. The sedimented cells were suspended in 200 μl of distilled water, and 100 μl of the cell suspension was plated on a minimal medium lacking leucine and cultured at 30°C for 2–3 days, after which the number of transformants on the medium was counted [66].

Figure 2 shows changes in the number of transformants and cell survival as a function of the initial intensity of electric field pulse. When the initial intensity of the single pulse was increased from 0 to 10 kV/cm at a fixed capacitance of 1 μF, the number of surviving cells decreased while the number of transformants (from *LEU2*$^-$ to *Leu2*$^+$) increased (see Fig. 2). Under these conditions, the number of transformants peaked at an initial intensity of 6.25 kV/cm, where the surviving cells were about 40% of the control. Essentially the same relationship was obtained between the number of surviving cells and that of transformants with a 2-μF or 4-μF pulse [66]. Note that without application of electric pulses, practically no transformants were obtained with the yeast strain, KK4, used here (see Fig. 2).

We also optimized the electrical conditions by changing capacitance of the discharge capacitor under fixed initial field intensity and by applying multipulses under fixed initial-field intensity and capacitance. The maximum number of transformants thus obtained was 90 ± 20/μg DNA by three

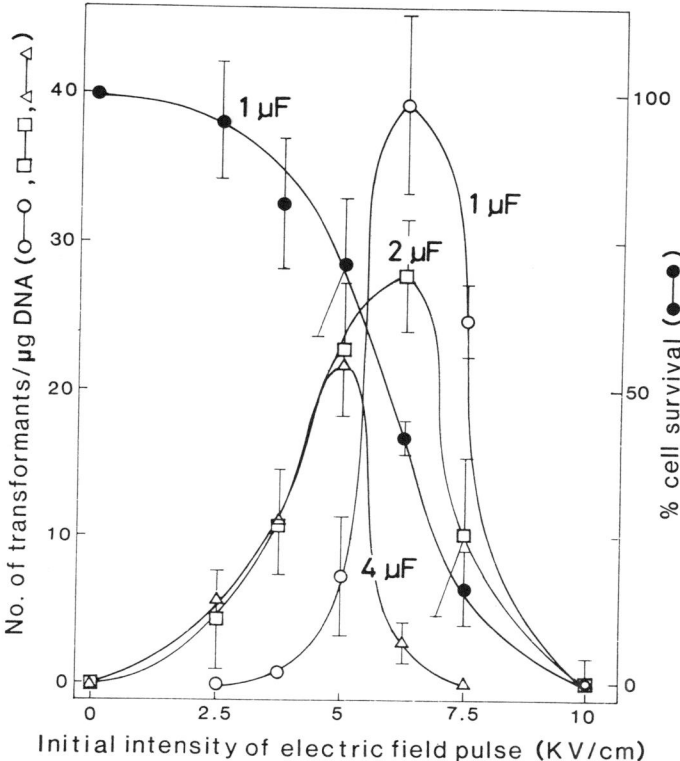

Fig. 2. Percent cell survival and number of transformants per μg DNA plotted against the initial intensity of the electric field pulse. The single pulses were applied by a capacitor-discharge method with a capacitance of 1 μF, 2 μF, or 4 μF. The bar represents the standard deviations of three experiments. (Reproduced from Hashimoto et al. [66] with permission.)

successive pulses with an initial field intensity of 5 kV/cm and with a capacitance of 1 μF.

This is the first report that intact yeast cells can be transformed with plasmid DNA by electric field pulses. This result indicates that plasmid DNA can be injected into cells through cell walls and membranes by electroinjection.

D. Electroinjection of Viral RNA Into Free Cells of Tobacco

Free mesophyll cells were enzymatically isolated from the leaves of *Nicotiana tabacum* cv. Samsun. They were then washed three times with a chamber medium containing 1.3 mM $CaCl_2$, 0.7 mM $ZnSO_4$, 0.5 mM $MgSO_4$, and 400 mM glucose. Tobacco mosaic virus (TMV) RNA was dissolved in distilled water and added to the cell suspension at a concentration of 10 µg RNA/ml, with 1×10^6 cells/ml. Cells + RNA were transferred to an electroinjection chamber, and electric field pulses were applied essentially as described. After application of the pulse, the cells were returned to a culture medium and cultured as described [67]. The electroinjected cells began to divide after about 6 days of culture. After 20 days of culture, small cell colonies composed of 10–50 cells had formed.

The electroinjected cells were assayed by an immunofluorescence technique with TMV fluorescent antibody. After 3–4 days of culture, some cells showed fluorescent specks owing to aggregates of TMV. When the cells were cultured for more than 20 days some had inclusion-body-like structures (arrows in Fig. 3A) in their cytoplasm. These "inclusion bodies" reacted with TMV antibody and fluoresced yellow-green (arrows in Fig. 3B).

Table IV summarizes the frequency of transfection in tobacco cells after electroinjection as assayed by an immunofluorescence technique. Clearly, as many as 50% of the cells that survived 24 days expressed genetic information encoded on the injected TMV RNA. No fluorescent bodies were detected by antibody staining of the control experiment specimens, in which neither TMV RNA nor an electric pulse treatment had been given to the cells. Furthermore, virus infectivity assay of the homogenate of the electroinjected cells showed local lesion on *N. glutinosa* leaves, evidence that the gene for virus multiplication also was expressed in the electroinjected cells [67]. Nishiguchi et al. [21] also have succeeded in the introduction and expression of TMV RNA in tobacco-free cells by electroinjection using square pulses.

The cell wall has been thought to be a major obstacle to DNA uptake into plant cells. The success of electroinjection of DNA or RNA into intact yeast cells or free tobacco cells, however, indicates that plant cells can take up genetic materials, even in the presence of cell walls, when an electric field pulse was applied to a cell-DNA (or -RNA) mixture. In order to quantify the effects of the presence of the cell walls upon the uptake of DNA into the cells, we recently examined the effects of cell wall formation upon transformation efficiency of tobacco (*N. tabacum* L. cv. Bright Yellow 2) protoplasts with plasmid DNA (35S-CAT-NOS [65]). After 1 day of protoplast culture, more than 90% of the cells formed cell walls, some divided after 3 days of the culture, and small cell colonies were formed by 7 days of the culture. Protoplast-derived cells (or walled protoplasts) were harvested after culture for 1 to 7 days, and the chimeric plasmid DNA was electroinjected into them. The CAT activity has been definitely detected with the electroinjected protoplast-derived cells. This result indicates that regen-

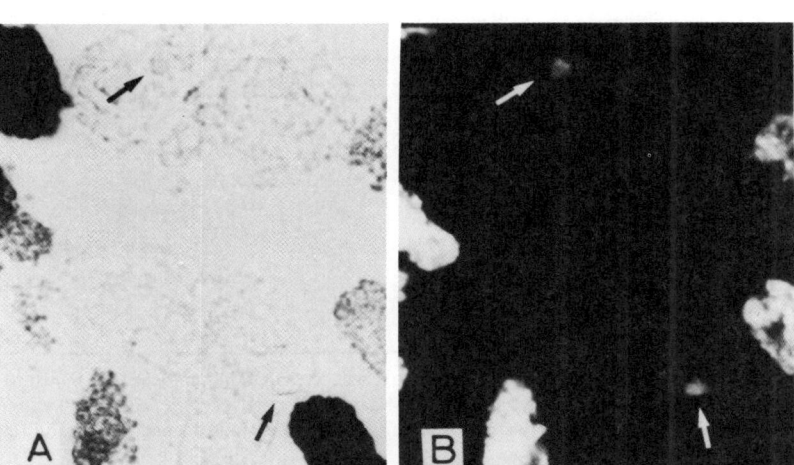

Fig. 3. Micrographs of mesophyll cells of tobacco (*Nicotiana tabacum* cv. Samsun) after electroinjection of TMV RNA. Lightfield (**A**) and darkfield (**B**) under epi-illumination with blue excitation light. Some cells contain inclusion-body-like structures (arrows in A) in their cytoplasm. These "inclusion bodies" have reacted with the TMV antibody and fluoresced yellow-green (arrows in B). Black patches in A represent the agar medium containing activated carbon that had been added to the culture medium. The strong spontaneous fluorescence in B is that of cells that did not survive the 24 days of culture. Electroinjection was carried out by a pulse (initial intensity of the electric field = 0.65 kV/cm with $\tau = 4$ μs) in the presence of 10 μg/ml TMV RNA. The electroinjected cells were cultured for 24 days in the dark, then assayed for the immunofluorescence with fluorescent TMV antibody. (Reproduced from Morikawa et al. [67] with permission.)

erated cell walls do not obstruct the electric-field-mediated entry of the plasmid DNA into the cells by electroinjection (Iida, Morikawa, and Yamada, in preparation). Further studies are currently being conducted.

In some plant species the ability of the plant cells to regenerate whole plants (totipotency) is enormously reduced when cell walls are digested to liberate protoplasts [88]. Thus we believe that development of electroinjection methods, i.e., creation of a direct transformation method for introducing genetic materials into intact ("walled") cells or tissues while keeping the totipotency of the cells intact, is vital for plant gene engineering. Pertinent experiments along this line currently are in progress.

ACKNOWLEDGMENTS

We are indebted to Professor D.E. Fosket for his critical reading of the manuscript and correcting our English. This work was in part supported by

TABLE IV. Frequency of Transfection (FT) Determined by Immunofluorescence Technique*

	Pulse condition		
Experiment No.	kV/cm	μF	FT[2]
Control(+ RNA)	0	0	0/30
Control(− RNA)	2.50	0.001	0/30
5161-1	0.65	0.002	33/66
5161-2	0.65	0.002	15/38
5161-3	0.65	0.002	45/168
5162-1	1.36	0.002	0/22
5162-2	1.36	0.002	8/29
5162-3	1.36	0.002	26/162

*TMV RNA (10 μg/ml) was electroinjected into tobacco (*Nicotiana tabacum* cv. Samsun) free mesophyll cells which then were cultured for 24 days. Only the surviving cells were examined because those that did not survive showed strong spontaneous fluorescence; but no fluorescent TMV bodies were present (see Fig. 3).
[a]FT = (the No. of cells containing fluorescent bodies)/(the No. of cells observed). (Reproduced from Morikawa et al. [67] with permission.)

Grants-in-Aid from the Ministry of Education, Science, and Culture of Japan.

REFERENCES

1. Senda M, Takeda J, Abe S, Nakamura T: Induction of cell fusion of plant protoplasts by electrical stimulation. Plant Cell Physiol 20:1441, 1979.
2. Neumann E, Gerisch G, Opatz K: Cell fusion induced by high electric impulses applied to *Dictyostelium*. Naturwissenschaften 67:414, 1980.
3. Weber H, Förster W, Jacob H-E, Berg H: Microbiological implications of electric field effects. III. Stimulation of yeast protoplast fusion by electric field pulses. Z Allg Mikrobiol 21:555, 1981.
4. Zimmermann U, Scheurich P: High frequency fusion of plant protoplasts by electric fields. Planta 151:26, 1981.
5. Watts JW, King JM: A simple method for large-scale electrofusion and culture of plant protoplasts. Biosci Rep 4:335, 1984.
6. Halfmann HJ, Emeis CC, Zimmermann U: Electro-fusion of haploid *Saccharomyces* yeast cells of identical mating type. Arch Microbiol 134:1, 1983.
7. Schnettler R, Zimmermann U, Emeis CC: Large-scale production of yeast hybrids by electrofusion. FEMS Microbiol Lett 24:81, 1984.
8. Watts JW, Doonan JH, Cove DJ, King JM: Production of somatic hybrids of moss by electrofusion. Mol Gen Genet 199:349, 1985.
9. Lo MMS, Tsong TY, Conrad MK, Strittmatter SM, Hester LD, Snyder SH: Monoclonal antibody production by receptor-mediated electrically induced cell fusion. Nature 310:792, 1984.

10. Morikawa H, Sugino K, Hayashi Y, Takeda J, Senda M, Hirai A, Yamada Y: Interspecific plant hybridization by electrofusion in *Nicotiana.* Bio/Technol 4:57, 1986.
11. Yamada Y, Morikawa H, Sato F, Yamamoto Y: Secondary plant products from cultured hybrid cells. Proc Jpn Acad Ser B 63:208, 1987.
12. Chapel M, Teissié J, Alibert G: Electrofusion of spermine-treated plant protoplasts. FEBS Lett 173:331, 1984.
13. Chapel M, Montane M-H, Ranty B, Teissié J, Alibert G: Viable somatic hybrids are obtained by direct current electrofusion of chemically aggregated plant protoplasts. FEBS Lett 196:79, 1986.
14. Sale AJH, Hamilton WA: Effects of high electric fields on micro-organisms. III. Lysis of erythrocytes and protoplasts. Biochim Biophys Acta 163:37, 1968.
15. Zimmerman U, Scheurich P, Pilwat G, Benz R: Cells with manipulated functions:new perspectives for cell biology, medicine, and technology. Angew Chem Int Ed Engl 20:325, 1981.
16. Mehrle W, Zimmermann U, Hampp R: Evidence for asymmetrical uptake of fluorescent dyes through electro-permeabilized membranes of *Avena* mesophyll protoplasts. FEBS Lett 185:89, 1985.
17. Sowers AE: Characterization of electric field-induced fusion in erythrocyte ghost membranes. J Cell Biol 99:1989, 1984.
18. Sowers AE, Lieber MR: Electropore diameters, lifetimes, numbers, and locations in individual erythrocyte ghosts. FEBS Lett 205:179, 1986.
19. Okada K, Nagata T, Takebe I: Introduction of functional RNA into plant protoplasts by electroporation. Plant Cell Physiol 27:258, 1986.
20. Watts JW, King JM, Stacey NJ: Inoculation of protoplasts with viruses by electroporation. Virology 157:40, 1987.
21. Nishiguchi M, Sato T, Motoyoshi F: An improved method for electroporation in plant protoplasts: Infection of tobacco protoplasts by tobacco mosaic virus particles. Plant Cell Rep 6:90, 1987.
22. Tempelaar MJ, Jones MGK: Fusion characteristics of plant protoplasts in electric fields. Planta 165:205, 1985.
23. Zachrisson A, Bornman CH: Electromanipulation of plant protoplasts. Physiol Plant 67:507, 1986.
24. Bates GW, Hasenkampf CA: Culture of plant somatic hybrids following electrical fusion. Theor Appl Genet 70:227, 1985.
25. Yamada Y, Morikawa H: Protoplast fusion of secondary metabolite-producing cells. In Neumann K-H, Barz W, Reinhard E (eds): "Primary and Secondary Metabolism of Plant Cell Cultures." Berlin: Springer-Verlag, 1985, p 255.
26. Tur-Kaspa R, Teicher L, Levine BJ, Skoultchi AI, Shafritz DA: Use of electroporation to introduce biologically active foreign genes into primary rat hepatocytes. Mol Cell Biol 6:716, 1986.
27. Austin S, Baer MA, Helgeson JP: Transfer of resistance of potato leaf roll virus from *Solanum brevidens* into *Solanum tuberosum* by somatic fusion. Plant Sci 39:75, 1985.
28. Gleddie S, Fassuliotis G, Keller WA, Setterfield G: Somatic hybridization as a potential method of transferring nematode and mite resistance into eggplant. Z Pflanzenzüchtg 94:352, 1985.
29. Bates GW, Gaynor JJ, Shekhawat NS: Fusion of plant protoplasts by electric fields. Plant Physiol 72:1110, 1983.
30. Kohn H, Schieder R, Schieder O: Somatic hybrids in tobacco mediated by electrofusion. Plant Sci 38:121, 1985.

31. Puite KJ, van Wikselaar P, Verhoeven H: Electrofusion, a simple and reproducible technique in somatic hybridization of *Nicotiana plumbaginifolia* mutants. Plant Cell Rep 4:274, 1985.

32. Morikawa H, Kumashiro T, Kusakari K, Iida A, Hirai A, Yamada Y: Interspecific hybrid plant formation by electrofusion in *Nicotiana*. Theor Appl Genet 75:1, 1987.

33. Puite KJ, Roest S, Pijnacker LP: Somatic hybrid potato plants after electrofusion of diploid *Solanum tuberosum* and *Solanum phureja*. Plant Cell Rep 5:262, 1986.

34. Pijnacker LP, Ferwerda MA, Puite KJ, Roest S: Elimination of *Solanum phureja* nucleolar chromosomes in *S. tuberosum* + *S. phureja* somatic hybrids. Theor Appl Genet 73:878, 1987.

35. de Vries SE, Jacobsen E, Jones MGK, Loonen AEHM, Tempelaar MJ, Wijbrandi J, Feenstra WJ: Somatic hybridization of amino acid analogue-resistant cell lines of potato (*Solanum tuberosum* L.) by electrofusion. Theor Appl Genet 73:451, 1987.

36. Richter H-P, Scheurich P, Zimmermann U: Electric field-induced fusion of sea urchin eggs. Dev Growth Differ 23:479, 1981.

37. Tempelaar MJ, Duyst A, De Vlas SY, Krol G, Symmonds C, Jones MGK: Modulation and direction of the electrofusion response in plant protoplasts. Plant Sci 48:99, 1987.

38. Bates GW: Electrical fusion for optimal formation of protoplast heterokaryons in *Nicotiana*. Planta 165:217, 1985.

39. Hibi T, Kano H, Sugiura M, Kazami T, Kimura S: High efficiency electro-transfection of tobacco mesophyll protoplasts with tobacco mosaic virus RNA. J Gen Virol 67:2037, 1986.

40. Morikawa H, Hayashi Y, Yamada Y: Manual pair formation and electrofusion of protoplasts using platinum microelectrodes. Plant Tissue Culture Lett 4 (in press).

41. Morikawa H, Hayashi Y, Hirabayashi Y, Asada M, Yamada Y: Cellular and vacuolar fusion of protoplasts electrofused using platinum microelectrodes. Plant Cell Physiol 29:189, 1988.

42. Vienken J, Zimmermann U: Electric field-induced fusion: Electro-hydraulic procedure for production of heterokaryon cells in high yield. FEBS Lett 137:11, 1982.

43. Tempelaar MJ, Jones MGK: Directed electrofusion between protoplasts with different responses in a mass fusion system. Plant Cell Rep 4:92, 1985.

44. Hampp R, Steingraber M: Electric-field-induced fusion of evacuolated mesophyll protoplasts of oat. Naturwissenschaften 72:91, 1985.

45. Naton B, Mehrle W, Hampp R, Zimmermann U: Mass electrofusion and mass selection of functional hybrids from vacuolate × evacuolate protoplasts. Plant Cell Rep 5:419, 1986.

45a. Morikawa H, Asada M, Yamada Y: Effect of physical parameters upon protoplast electrofusion. Plant Cell Physiol 29 (in press).

46. Pilwat G, Richter H-P, Zimmerman U: Giant culture cells by electric field-induced fusion. FEBS Lett 133:169, 1981.

47. Ohno-Shosaku T, Okada Y: Facilitation of electrofusion of mouse lymphoma cells by the proteolytic action of proteases. Biochem Biophys Res Commun 120:138, 1984.

48. Ruzin SE, McCarthy SC: The effect of chemical facilitators on the frequency of electrofusion of tobacco mesophyll protoplast. Plant Cell Rep 5:342, 1986.

49. Abe S, Takeda J: Possible involvement of calmodulin and the cytoskeleton in electrofusion of plant protoplasts. Plant Physiol 81:1151, 1986.

50. Terada R, Yamashita Y, Nishibayashi S, Shimamoto K: Somatic hybrids between *Brassica oleracea* and *B. campestris:* Selection by the use of iodoacetamide inactivation and regeneration ability. Theor Appl Genet 73:379, 1987.

51. Auer D, Brandner G, Bodemer W: Dielectric breakdown of the red blood cell membrane and uptake of SV40 DNA and mammalian cell RNA. Naturwissenschaften 63:391, 1976.

52. Neumann E, Schaefer-Ridder M, Wang Y, Hofschneider PH: Gene transfer into mouse lyoma cells by electroporation in high electric fields. EMBO J 1:841, 1982.

53. Davey MR, Cocking EC, Freeman J, Pearce N, Tudor I: Transformation of *Petunia* protoplasts by isolated *Agrobacterium* plasmids. Plant Sci Lett 18:307, 1980.

54. Krens FA, Molxendijk L, Wullems GJ, Schilperoort RA: *In vitro* transformation of plant protoplasts with Ti-plasmid DNA. Nature 296:72, 1982.

55. Paszkowski J, Shillito RD, Saul M, Mandák V, Hohn T, Hohn B, Potrykus I: Direct gene transfer to plants. EMBO J 3:2717, 1984.

56. Hain R, Stabel P, Czernilofsky AP, Steinbiss HH, Herrera-Estrella L, Schell J: Uptake, integration, expression and genetic transmission of a selectable chimaeric gene by plant protoplasts. Mol Gen Genet 199:161, 1985.

57. Potrykus I, Paszkowski J, Saul MW, Petruska J, Shillito RD: Molecular and general genetics of a hybrid foreign gene introduced into tobacco by direct gene transfer. Mol Gen Genet 199:169, 1985.

58. Lörz H, Baker B, Schell J: Gene transfer to cereal cells mediated by protoplast transformation. Mol Gen Genet 199:178, 1985.

59. Potrykus I, Saul MW, Petruska J, Paszkowski J, Shillito RD: Direct gene transfer to cells of a graminaceous monocot. Mol Gen Genet 199:183, 1985.

60. Balázs E, Bouzoubaa S, Guilley H, Jonard G, Paszkowski J, Richards K: Chimeric vector construction for higher-plant transformation. Gene 40:343, 1985.

61. Werr W, Lörz H: Transient gene expression in a Gramineae cell line. A rapid procedure for studying plant promoters. Mol Gen Genet 202:471, 1986.

62. Paszkowski J, Pisan B, Shillito RD, Hohn T, Hohn B, Potrykus I: Genetic transformation of *Brassica campestris var. rapa* protoplasts with an engineered cauliflower mosaic virus genome. Plant Mol Biol 6:303, 1986.

63. Uchimiya H, Fushimi T, Hashimoto H, Harada H, Syono K, Sugawara Y: Expression of a foreign gene in callus derived from DNA-treated protoplasts of rice (*Oryza sativa* L.). Mol Gen Genet 204:204, 1986.

64. Fromm M, Taylor LP, Walbot V: Expression of genes transferred into monocot and dicot plant cells by electroporation. Proc Natl Acad Sci USA 82:5824, 1985.

65. Fromm ME, Taylor LP, Walbot V: Stable transformation of maize after gene transfer by electroporation. Nature 319:791, 1986.

66. Hashimoto H, Morikawa H, Yamada Y, Kimura A: A novel method for transformation of intact yeast cells by electroinjection of plasmid DNA. Appl Microbiol Biotechnol 21:336, 1985.

67. Morikawa H, Iida A, Matsui C, Ikegami M, Yamada Y: Gene transfer into intact plant cells by electroinjection through cell walls and membranes. Gene 41:121, 1986.

68. Köhler F, Golz C, Eapen S, Kohn H, Schieder O: Stable transformation of moth bean *Vigna aconitifolia* via direct gene transfer. Plant Cell Rep 6:313, 1987.

69. Ecker JR, Davis RW: Inhibition of gene expression in plant cells by expression of antisense RNA. Proc Natl Acad Sci USA 83:5372, 1986.

70. Ou-Lee TM, Turgeon R, Wu R: Expression of a foreign gene linked to either a plant-virus or a *Drosophila* promoter, after electroporation of protoplasts of rice, wheat, and sorghum. Proc Natl Acad Sci USA 83:6815, 1986.

71. Howard EA, Walker JC, Dennis ES, Peacock WJ: Regulated expression of an alcohol dehydrogenase 1 chimeric gene introduced into maize protoplasts. Planta 170:535, 1987.

72. Okada K, Takebe I, Nagata T: Expression and integration of genes introduced into highly synchronized plant protoplasts. Mol Gen Genet 205:398, 1986.
73. Langridge WHR, Li BJ, Szalay AA: Electric field mediated stable transformation of carrot protoplasts with naked DNA. Plant Cell Rep 4:355, 1985.
74. Shillito RD, Saul MW, Paszkowski J, Müller M, Potrykus I: High efficiency direct gene transfer to plants. Bio/Technol 3:1099, 1985.
75. Negrutiu I, Shillito R, Potrykus I, Biasini G, Sala F: Hybrid genes in the analysis of transformation conditions. Plant Mol Biol 8:363, 1987.
76. Riggs CD, Bates GW: Stable transformation of tobacco by electroporation: Evidence for plasmid concatenation. Proc Natl Acad Sci USA 83:5602, 1986.
77. Christou P, Murphy JE, Swain WF: Stable transformation of soybean by electroporation and root formation from transformed callus. Proc Natl Acad Sci USA 84:3962, 1987.
78. Boston RS, Becwar MR, Ryan RD, Goldsbrough PB, Larkins BA, Hodges TK: Expression from heterologous promoters in electroporated carrot protoplasts. Plant Physiol 83:742, 1987.
79. Gusew N, Nepveu A, Chartrand P: Linear DNA must have free ends to transform rat cells efficiently. Mol Gen Genet 206:121, 1987.
80. Peerbolte R, Krens FA, Mans RMW, Floor M, Hoge JHC, Wullems GJ, Schilperoort RA: Transformation of plant protoplasts with DNA: Cotransformation of non-selected calf thymus carrier DNA and meiotic segregation of transforming DNA sequences. Plant Mol Biol 5:235, 1985.
81. Watanabe Y, Meshi T, Okada Y: Infection of tobacco protoplasts with in vitro transcribed tobacco mosaic virus RNA using an improved electroporation method. FEBS Lett 219:65, 1987.
82. Nishiguchi M, Langridge WHR, Szalay AA, Zaitlin M: Electroporation-mediated infection of tobacco leaf protoplasts with tobacco mosaic virus RNA and cucumber mosaic virus RNA. Plant Cell Rep 5:57, 1986.
83. Newport JW, Kirschner MW: Regulation of the cell cycle during early Xenopus development. Cell 37:731, 1984.
84. Benz R, Zimmermann U: The resealing process of lipid bilayers after reversible electrical breakdown. Biochim Biophys Acta 640:169, 1981.
85. Senda M, Morikawa H, Takeda J: Electrical induction of cell fusion of plant protoplasts. In Fujiwara A (ed): "Plant Tissue Culture 1982." Tokyo: Maruzen, 1982, p 615.
86. Karube I, Tamiya E, Matsuoka H: Transformation of *Saccharomyces cerevisiae* spheroplasts by high electric pulse. FEBS Lett 182:90, 1985.
87. Hauptmann RM, Ozias-Akins P, Vasil V, Tabaeizadeh Z, Rogers SG, Horsch RB, Vasil IK, Fraley RT: Transient expression of electroporated DNA in monocotyledonous and dicotyledonous species. Plant Cell Rep 6:265, 1987.
88. Vasil V, Vasil IK: Isolation and culture of embryogenic protoplasts of cereals and grasses. In Vasil IK (ed): "Cell Culture and Somatic Cell Genetics of Plants." London: Academic, 1984, p 398.

Biotechnology in Agriculture, pages 203–223
© 1988 Alan R. Liss, Inc.

Applications of Somaclonal Variation

David A. Evans

DNA Plant Technology Corporation, Cinnaminson, New Jersey 08077

———◆◆———

———◆◆———

The ability to recover intact plants from cultured cells was originally viewed as an efficient method of producing large numbers of clones, an expectation that has been realized. Many commercial laboratories now use tissue culture of shoot tips to propagate a wide range of ornamental plants, including orchids, ferns, lilies, etc., and some crop plants, such as oil palm. As predicted, tissue culture propagation has been shown to be economically competitive with conventional propagation and has resulted to a large extent in clonal fidelity [1]. However, it was recognized in early experiments that if cultures were established from explants that did not contain a pre-organized meristem, or if cultures were maintained as callus prior to plant regeneration, the regenerated plants were quite variable. In early reports most of the variation was attributed to the readily detected chromosomal instability of cultured plant cells. In many of these studies, the degree of chromosome instability was reported to be proportional to the length of time the cells remained in culture. Recognition of the spontaneous variation inherent in long-term culture led to the use of cell culture for mutagenesis and selection of genetic variants and for direct recovery of novel genotypes from cell cultures via somaclonal variation (Fig. 1). In this paper, we intend to outline

Fig. 1. Schematic of procedure for producing somaclonal variation.

the types of genetic variants that have been recovered from cell cultures via somaclonal variation and to discuss the impact of this variation for crop improvement and recovery of new ornamentals.

I. TERMINOLOGY

As genetic variation has been detected in plants for several years, it is not surprising that both general cell culture and specific genetic terminology have evolved that describe the variation recovered from cultured cells. Alterations have been referred to as phenotypic or genotypic changes. The genotype refers to the sum total of the genetic information, while the phenotype, or appearance of a plant or cell, is recognized to be a combination of genetic (genotypic) and environmental factors. These phenotypic changes that are not the result of genetic alterations are termed epigenetic changes. Hence, in most cases, it is appropriate to characterize variation in the plant or plant cell culture phenotype as a genetic or an epigenetic change. The distinction between these two types of changes is only conclusively demonstrated by detailed genetic evaluation often requiring several sexual generations. Because of this requirement, very few reports of cell-culture-induced variation have been conclusively shown to have a genetic basis.

Early variant plants regenerated from cell cultures of geranium were termed "callicones" [2], while plants regenerated from protoplasts of potato were termed "protoclones" [3]. Larkin and Scowcroft [4] promoted the use of the more general term "somaclonal variation" for the variation detected in plants derived from any form of cell culture. However, the type of genetic variation recovered in regenerated plants is to a large extent dictated by the genetic constitution of the particular cell population that is regenerated. It is necessary for genetic reasons to distinguish between plants regenerated from somatic and gametic tissue. To this end we use the term "somaclones" to refer to plants regenerated from cell cultures originating from somatic tissue and "gametoclones" to refer to plants regenerated from cell cultures originating from gametic tissue.

Similarly, mixed terminology has evolved to refer to sexual progeny of plants regenerated from cell cultures. In this paper, as in our earlier publications [5,6], we use the terminology proposed by Chaleff [7, pp. 94–99] as follows: Plants regenerated from cell culture, irrespective of tissue or origin, are referred to as R or R_0 plants. The self-fertilized progeny of R plants are referred to as R_1 plants. Subsequent generations produced by self-fertilization are termed R_2, R_3, R_4, etc.

II. REVIEW OF LITERATURE

Although cloning of identical plants was originally believed to be the principal use for plant cell culture, it has become increasingly clear that under the appropriate cultural conditions, a great deal of genetic variability can be recovered in regenerated plants (Table I). Results from early experiments documented genetic variability in plants regenerated from protoplasts [3]. Growth of unorganized callus was thought to be necessary for induction of variability [2], but recent results suggest that genetic variability is even present in population of plants regenerated directly from leaf explants in the absence of callus growth. Phenotypic variation has been reported in a number of plant species regenerated via organogenesis or embryogenesis.

Genetic variation was first detected as altered chromosome numbers in cultured plant cells. The onset of chromosome instability has been well characterized in many systems, such as *Daucus carota* [17] and *Haplopappus gracilis* [18]. The most frequently reported variation has been polyploidy [19], attributed to selective growth of normally nondividing polyploid cells preexisting in the original explant [20]. It has also been

TABLE I. Documented Cases of Somaclonal Variation in Crop and Ornamental Species

Species	Altered characteristics	Inheritance	Reference
Apium graveolens (celery)	Isozyme variation; chromosome structure	No	[8]
Chrysanthemum x *morifolium*	Flower size and color, petal shape	No	[9]
Dendrobium	Flower size and color, petal shape	No	[10]
Lactuca sativa (lettuce)	Plant height and vigor, fertility, leaf color, chlorophyll pigment	Yes	[11]
Lycopersicon esculentum (tomato)	Fruit color, growth habit, fertility, flower color, chlorophyll content	Yes	[5]
Medicago sativa (alfalfa)	Yield, fertility	Yes	[12]
Nicotiana alata	Time to flower, flower and leaf morphology, pollen viability, plant height	Yes	[13]
Pelargonium spp.	Plant and organ size, leaf and flower morphology, oil constituents, fasciation, pubescence, anthocyanin pigmentation	No	[2]
Saintpaulia ionantha	RuBCase activity	No	[14]
Solanum tuberosum (potato)	Chromosome structure, yield, disease resistance	No	[15]
Zea mays (maize)	Early tasseling, leaf spotting, curled leaves, vigor, tillering	Yes	[16]

reported that the frequency of polyploid cells is dependent on the concentration and type of cytokinin used in the culture medium. Bennici et al. [21] reported that the frequency of polyploid cells in cultures of *H. gracilis* was dependent on the ratio of kinetin to naphthaleneacetic acid (NAA). Polyploid plants have been recovered in many commercially important plant species including *Pelargonium zonale* [2], *Nicotiana tabacum* [22], *N. alata* [13], *Lycopersicon esculentum* [5], and *Medicago sativa* [23]. Aneuploid changes, typically involving the gain or loss of a few chromosomes, have also been frequently reported in plant cell cultures. The accumulation of aneuploid cells has been attributed to aging of cultures [20], and in many cases, older cell lines are incapable of plant regeneration. However, aneuploid plants have been regenerated from cultures of *N. tabacum* [24] and *Saccharum* spp. [25]. Aneuploidy is often associated with sterility but would certainly be tolerated in plant species that are propagated asexually. In addition to these numerical chromosome changes, observation of anaphase bridges and fragments also suggests that the structure of plant chromosomes is modified in cell culture [18]. Chromosome rearrangements have been detected in clones of potatoes (*Solanum tuberosum*) regenerated from mesophyll protoplasts [15].

Evidence of recovery of single gene mutations produced via somaclonal variation has been presented. The progeny of tomato plants regenerated from leaf-derived callus were examined [5] and 13 distinct single gene mutations were recovered among 230 regenerated tomato plants. This frequency of visual somaclonal mutations (ca. 1 in 18 regenerated plants) is substantially greater than the cell mutagenesis rate from several cell selection experiments [26]. Several of these single gene mutants of tomato have been well characterized and mapped to specific chromosomes [27].

Cytoplasmic genetic variation has been detected less frequently following plant regeneration from cell culture, although the importance of this class of variation has been emphasized [28]. The most compelling evidence for cytoplasmic genetic variation among regenerated plants was presented by Gengenbach et al. [29], who used restriction enzyme analysis of isolated mitochondrial DNA to detect spontaneous variant corn plants regenerated from cell culture. These authors monitored variation in cytoplasmically controlled male sterility, a mitochondrially encoded character. Hence, somaclonal variation has resulted in numerical and structural chromosomal changes, in nuclear genetic modifications and in cytoplasmic genetic variation. This broad spectrum of variation suggests that by using appropriate selection methods all classes of genetic variation could be recovered and used for crop improvement.

No detailed systematic studies have been devised to ascertain the source of somaclonal variation; however, it is possible by examining published literature to gain information regarding the source of variation. Skirvin and Janick [2] systematically compared plants regenerated from callus of five cultivars of *Pelargonium* spp. Plants obtained from geranium stem cuttings in vitro were uniform, whereas plants from in vivo root and petiole cuttings and plants regenerated from callus were quite variable. This suggests that some variability is correlated with donor explant and preexists in the tissue used to establish cell cultures. On the other hand, evidence from several laboratories suggests that variability is dependent on hormone concentration of the culture medium. Most likely variability is the result of both preexisting genetic instability and genetic alterations induced during the process of cell culture and subsequent plant regeneration.

Recent indications of somaclonal variation in several crop plants have stimulated interest in application of this method for crop improvement. Initial studies with sugar cane suggested that clones with disease resistance could be regenerated from callus induced from sensitive plants. Krishnamurthi and

Tlaskal [30] isolated clones of regenerated sugar cane with resistance to Fiji disease virus and downy mildew. These selected resistant clones did not show reduced sucrose yield. Similarly variants of sugar cane have been isolated with resistance to eyespot disease [4]. Most sugar cane variants have been attributed to changes in chromosome number. Researchers in several laboratories have examined plants regenerated from protoplasts of potato following the recovery of somaclones resistant to late blight [31]. Potato variants have been isolated with resistance to early blight and with altered growth habit, tuber shape and color, and maturity date. These somaclonal variants of potato were attributed to changes in chromosome number [32] and chromosome structure [15]. While these variants can be stably propagated asexually, the genetic inheritance of this variation in sugar cane and potato has not been determined. Variation with potential agricultural application has been detected to several plant species that are sexually propagated. These include yield improvement in tobacco, tobacco mosaic virus (TMV) resistance in tobacco, southern leaf blight (*Helminthosporium maydis*) resistance in corn (*Zea mays* L.), and leaf shape in geranium [27]. Somaclonal variants have also been described in rapeseed for blackleg disease susceptibility [33] and for growth habit, fruit color, and male sterility in tomato [5]. All of these variants demonstrate the potential of somaclonal variation for production of new breeding lines.

As detailed genetic analysis have been completed for only a few somaclonal variants, it is not surprising that this method has not been exploited. Most variants reported to date have been chromosomal. While chromosomal changes can be tolerated in asexually propagated crops such as potato and sugar cane, other types of single nuclear gene and cytoplasmic variation may be much more important for crop improvement of sexually propagated crops.

III. TOMATO IMPROVEMENT

Evans and Sharp [5] and Evans et al. [6] published a description of experiments designed to generate somaclones of tomato and to ascertain the genetic basis of somaclonal variation (Fig. 1).

Tomato seeds of a well-characterized open-pollinated tomato variety, UC82B, were sown in a greenhouse. Young, fully expanded leaves were removed from plants, sterilized, and placed onto a culture medium known to permit regeneration of plants from cultured leaf explants [34]. Callus that developed from the cultured leaf explants regenerated shoots within 4 weeks. Regenerated shoots were rooted on medium containing naphthalene acetic acid (NAA), then transferred to the greenhouse for maturation and fruit

Fig. 2. Male sterile somaclonal variant in tomato. The mutation was found to be recessive.

collection. Self-fertilized seed was collected from mature fruits on R_0 plants. To evaluate the R_1 generation, seed was sown in the greenhouse where seedling characters, such as chlorophyll deficiency, anthocyanin content, and leaf shape were monitored. R_1 seedlings were transplanted to the field to classify mature plant characteristics such as pedicel type, fruit shape, and fruit and flower color. In addition, preliminary data on agronomic characters were collected in the R_1 field evaluation.

Variation in chromosome number, particularly tetraploid, $2n = 48$, was detected. Sterile aneuploid lines were also detected. However, since emphasis was directed to analysis of R_1 progeny of regenerated plants, most aneuploid lines were discarded because little or no R_1 seed was collected. Several R_1 progenies were observed to segregate for morphological characters. These include recessive mutations for male sterility (Fig. 2), jointless pedicel, tangerine virescent leaf, flower and fruit color, lethal chlorophyll deficiency, virescence, and mottled leaf appearance and dominant mutations controlling fruit ripening and growth habit. Genetic analysis was completed by evaluating self-fertilized R_2 progeny of selected R_1 plants, and in some cases, by crossing with known mutant lines. Extensive genetic analysis has been completed for several of the variants.

Genetic analysis was first completed for the tangerine-virescent (tv-tc1) character. This character is a single recessive allele that results in orange flowers and fruit and yellow virescent leaves. The R_1 segregation data observed in the field (30 red to 6 tangerine fruit) first suggested control of the trait by a single recessive allele. Fruit was collected from eight individual

TABLE II. Genetic Analysis of Tangerine-Virescent Somaclone

Progeny	Red	Orange	Normal	Virescent
R_1	30	6	30	6
R_2 from orange R_1	0	102	0	102
R_2 from homozygous red R_1	125	0	125	0
R_2 from heterozygous red R_1	58	17	58	17
Orange virescent $R_2 \times$ t/t	0	20	20	0
F_2 of tv-tc1/t	0	97	72	30

self-fertilized plants. Six of these R_1 plants had red fruit while two had tangerine fruit. The two tangerine fruit contained seed that bred true for the tangerine-virescent phenotype. Of the six red-fruited plants, three bred true for red fruit and three segregated for red vs. tangerine-virescent. The pooled progeny of the three segregating plants fit a 3:1 ratio (58 red to 17 tangerine). In addition, among the segregating R_2 progeny, the flower, fruit, and leaf color defects cosegregated suggesting control of two pigments, carotenoids (flower and fruit) and chlorophyll (leaf), by a single pleiotropic gene. Such pleiotropic genes controlling carotenoids and chlorophyll, two compounds both found in plastids, have been reported previously. A single mutant R_1 plant was crossed to a known fruit variant, tangerine, and all hybrid progeny had tangerine fruit. As these two mutants do not complement (t-gene), it can be concluded that the new somaclone (tv-tc1) was a mutant for a new allele in a previously known gene at position 95 on the long arm of chromosome 10. This has been the most precise genetic characterization of a new somaclonal variant to date. In addition, by evaluating the self-fertilized progeny of the hybrids between the tv-tc1 and the earlier tangerine mutant for the virescence character (Table II) it has been possible to conclude that the locus contains two elements, one controlling chlorophyll synthesis and one controlling carotenoid synthesis, that mutate independently. The new tv allele recovered by somaclonal variation is recessive to the t allele for the virescence phenotype.

A second fruit color mutant was detected in a somaclone of variety C40 (Campbell Soup Company). The original regenerated plant was red, and the self-fertilized progeny of this plant segregated in a 3:1 ratio (Table III) as based on chi-square test ($\chi^2 = 0.5977$, 1 d.f., $0.30 < P < 0.50$). This mutant had no apparent pleitropic effects when evaluated in field trials. When genetic analysis was completed (Table III), it was found to be allelic to and virtually indistinguishable from the r-2 mutant. This new somaclone is located on chromosome 3 at position 29 [27].

Jointless pedicel is a trait that is a well-characterized recessive mutant of

TABLE III. Genetic Analysis of Yellow-Fruited Somaclone

Progeny	Red	Yellow
R_1 generation	15	4
R_2 from yellow-fruited R_1	0	99
R_2 from homozygous red-fruited R_1	132	0
R_2 from heterozygous red-fruited R_1	227	68
Yellow-fruited R_2 × r-2/r-2	0	24
Yellow-fruited R_2 × t/t	15	0

TABLE IV. Genetic Analysis of Jointless Somaclone

Progeny	Jointed	Jointless
R_1	0	30
R_2	0	123
R_2 × j-1/j-1	23	2
R_2 × j-2/j-2	0	24

tomato. Plants with this character are desirable for mechanical harvesting as the harvested fruit has no stem attached. Two mutants with jointless pedicel were identified in somaclone of UC82B. The first was as expected, i.e., the regenerated plant was normal, and the recessive mutant was detected in about 25% of the R_1 progeny. However, the original regenerated plant of the second variant (j-tc2) already expressed the mutant trait. This jointless R_0 plant bred true in both the R_1 and R_2 generations (Table IV). Based on crosses with a known jointless mutant and with a normal jointed tomato, we have been able to ascertain that the original R_0 plants was homozygous recessive for the jointless mutant. Presumably this somaclone originated by mutation, followed by mitotic recombination and subsequent shoot regeneration. This new mutation complements the known j mutation so that it is not encoded in the same gene. However, it does not complement the known j-2 mutant, and is, therefore, allelic with this somaclonal mutant and maps at position 0 on the short arm of chromosome 11.

The mottled mutation (m-tc1) uncovered among somaclones of UC82B is a chlorophyll-deficient mutant that was identified in the R_1 generation in the greenhouse. Selected mottled R_1 plants breed true for the mottled appearance. The variegation is somewhat similar to the previously reported plastome mutants. When chlorophyll deficient sectors of plastome mutants are placed in vitro, it is possible to establish shoot cultures of pure chlorophyll-deficient tissue [35]. However, the mottled phenotype of m-tc1 is quite sable and is maintained in shoots regenerated from leaf explants of mottled plants. In addition, it is also possible to discern orange-red mottling

TABLE V. Field Performance of *Fusarium* 2 Resistant Somaclone (DNAP 17)

	Yield (kg/plot)	Soluble solids
New Jersey		
DNAP 17	154.4	4.70
UC82B	156.8	4.34
Mexico		
DNAP 17	140.1	4.36
UC82B	137.1	4.39

on fruits of homozygous (m-tc1/m-tc1) plants. Once again, as with tv-tc1, the m-tc1 mutant has an effect on both the chlorophyll and carotenoid pigments. This mutant appears to be distinct from any of the hundreds of previously reported tomato mutants.

Several other single gene mutations have now been identified in R_1 progeny tests of regenerated tomato plants, including a semidominant allele controlling an electrophoretic variant of alcohol dehydrogenase and a dominant allele conferring resistance to *Fusarium oxysporum* Race 2 [36]. In addition, several somaclones have been identified with agriculturally important traits that do not appear to be controlled by single genes. For example, new lines have been developed with higher pigment and higher solids.

The *Fusarium* Wilt 2 resistance somaclone is of interest as it is a single gene trait of agronomic importance. Most new tomato varieties contain the I-2 gene for *Fusarium* Wilt 2 resistance derived from *L. pimpinellifolium*. This new somaclone is allelic to the I-2 gene and is, therefore, located on chromosome 11 at position 74. Moreover, it appears to have retained most of the characteristics typical of its parent variety, UC82B (Table V). The somaclone does appear to have slightly softer fruits than UC82B, although the fruit would be suitable for mechanical harvesting. Soft fruit may represent a pleiotropic effect of this new somaclone but must be characterized in greater detail.

Morphological variants have also been detected among regenerated plants of UC82B that bred true in the R_1 generation. In many cases, the altered somaclones have larger leaf size, darker leaf color, or reduced fruit set. In some cases, these R_1 plants have been evaluated, and preliminary evidence obtained by restriction enzyme analysis of isolated chloroplast DNA suggests that genetic changes have occurred in chloroplast DNA [6]. Chloroplast genes are inherited maternally in tomato; hence, one would expect the variant to breed true in R_1 progeny. It is not surprising that chloroplast DNA-variants are uncovered as the number of plastids in a developing shoot apex is such smaller than in a mature cell [cf. 37]. Hence, if a mutation occurs in chloroplast DNA, it is more likely to become the dominant plastid type during sorting out if the mutant occurs in 1 of 10 plastids than in 1 of 100 plastids, for example.

In addition, several other tomato variants have been regenerated using the procedure developed for UC82B. Single gene mutations have been recovered in several of these varieties including several male sterile and chlorophyll deficiency mutation. A recessive mutation for chlorophyll deficiency was identified in one tomato breeding line. This new mutant appears distinct from other known chlorophyll-deficient mutations. New fruit color mutants have also been detected, including a dominant orange mutation in C38. C38 is a Campbell Soup Company commercial tomato variety. This orange mutation appears to be pleiotropic as it is associated with bushy foliage and altered flower color.

Based on our work, the following has been concluded regarding soma-clonal variation in tomato: 1) chromosome number variation can be recovered in regenerated plants; 2) several single gene mutations have been recovered in several different tomato varieties; 3) somaclones include dominant, semidominant, and recessive nuclear mutations; 4) the frequency of single gene mutation using our procedure is in the neighborhood of 1 mutant in every 20–25 regenerated plants; 5) some evidence suggests that new single gene mutants not previously reported using conventional muta-genesis have been recovered using somaclonal variation; 6) the occurrence of 3:1 ratios for single gene mutants in R_1 plants suggests that mutants are of clonal origin and that the mutation occurred prior to shoot regeneration (i.e., no mosaics are detected); 7) evidence suggests that mitotic recombination (reciprocal or nonreciprocal) may also account for some somaclonal varia-tion; 8) evidence suggests that mutation in chloroplast DNA (detected by both maternal inheritance and restriction enzyme analysis) can also be recovered; and 9) agriculturally useful variants leading to development of new breeding lines have been recovered via somaclonal variation.

IV. DEVELOPMENT OF NEW ORNAMENTALS

As demonstrated by the work on tomato, it is possible to use somaclonal variation to uncover new, useful genetic variants. In addition to those variants already described, many morphological variants, with novel leaf, flower, and fruit color were detected. These unique variants, though not useful for a commercial horticultural crop, are the type of variants often used to release new, ornamental varieties. To evaluate the potential value of somaclonal variation in the development of new ornamental varieties, we regenerated plants of *Nicotiana alata* "Nicki Red."

Nicotiana alata has a diploid chromosome number of 18. It is a member of the section Alatae and is self-incompatible. Young, fully expanded leaves of *Nicotiana alata* cv. Nicki Red (Park Seed Company, Greenwood, SC) were surface sterilized in 8% Clorox for 8 min, then rinsed three times in sterile distilled water. Leaves were dried under sterile conditions and then cut with a sterile scalpel into 2.3 × 2.5-cm sections and placed onto agar-solidified media composed of Murashige-Skoog macro- and micronutrients and B5 vitamins. This modified Murashige and Skoog (MMS) medium was supplemented with 0, 0.5, 1.0, 5.0, or 10.0 μM of 6-benzyladenine (BA) to determine capacity for plant regeneration.

Leaf explants were cultured under 16 h of light with a mean temperature of 29.4°C (85°F). When regenerated shoots had reached a height of 3 cm, they were excised and placed onto rooting medium: MMS with 0.1 mM 3-aminopyridine [38]. Following root development, shoots were transplanted to Jiffy pots and subsequently to 4-in pots when secondary leaves developed. Plants were transferred to the greenhouse, where they were raised to maturity.

Shoots were regenerated from explants cultured on either 1.0 μM or 5.0 μM BA within 28 days. After 56 days in culture, none of the explants on hormone-free medium had regenerated shoots. In medium containing 0.5 μM BA, a small amount of callus was formed from which only one small shoot was regenerated from the eight explants at this treatment. Large shoots with normal morphology were obtained using 1.0 μM BA, but the frequency of shoot formation and number of shoots recovered was less than on the medium containing either 5.0 μM or 10.0 μM BA. Five micromolar BA was optimum, as it resulted in the highest number of shoots in the shortest time period.

The regenerated plants were phenotypically distinct from seed-derived plants. One characteristic frequently observed in these regenerated plants was not heritable. Seventy-two percent of the plants regenerated in the presence of 10 μM BA exhibited first and second leaves that were wider than young leaves of seed grown *N. alata*. However, subsequent leaves of these regenerated plants developed normally. Hence, this trait is epigenetic. On the other hand, a small percentage of plants regenerated on MMS containing 5.0–10.0 μM BA flowered as soon as they were transferred to 4-in pots, at a height of only 3 in. This type of dwarfness was sexually transmitted to the R_1 self-fertilized progeny of these plants (Table VI). Plant height behaved as a quantitatively inherited character as a range of intermediate heights were observed among the R_1 progeny. While the population size was insufficient to ascertain the mode of inheritance, it is nonetheless clear that this trait expressed as mean plant height, is stably transmitted to the R_1 generation.

TABLE VI. Statistical Comparison of Dwarf *Nicotiana alata* Self-Fertilized Progeny for Flowering Heights (Leaf Explants Cultured on MS Medium Containing 5 or 10 μM BA)

Plant material	Mean height (cm)	t-test	Probability
Control (seed-derived)	52.0	—	—
10 μM BA regenerate, #3	29.5	2.52	0.01
10 μM BA regenerate, #5	38.7	2.45	0.025
10 μM BA regenerate, #10	37.2	1.97	0.05
Control (seed-derived)	42.3	—	—
5 μM BA regenerate, #21	28.7	2.53	0.01
5 μM BA regenerate, #23	20.6	4.19	0.01

Evaluation of the R_1 generation will permit a more detailed genetic analysis of this trait. Changes in gross chromosome number can be excluded from consideration as all regenerated dwarf plants had $2n = 18$.

Chromosome number was ascertained for each plant regenerated in the presence of 5.0 μM and 10.0 μM BA. At both of these cytokinin concentrations 25% tetraploidy was observed in the root tips of the regenerated plants. These tetraploid regenerates were much slower to flower than comparable diploid regenerates. After 6 months from the date of culture, all of the diploid regenerates, but less than 50% of the tetraploid regenerates, had flowered.

Some tetraploid regenerates were crossed with diploid regenerates. Tetraploid plants were used both as female and male. Crosses were completed using regenerates that had already been observed to self-pollinate and set seed. It was presumed that these plants would have a higher probability of successful crossing than regenerates that had been unable to set seed. Several of these crosses were successful; however, triploid progeny were recovered only when the tetraploid was used as the female parent.

Plants regenerated on MMS containing 5.0 or 10.0 μM BA exhibited decreased pollen variability compared to seed-derived plants when stained with acetocarmine. Plants regenerated on 10 μM BA were also observed to produce less pollen than normal *N. alata*. The pollen viability remained low even after plants were 7 mo old. These plants eventually produced a small amount of self-pollinated seed after 10–12 mo in the greenhouse.

The flower morphology of several regenerated plants was also altered (Fig. 3). This trait has not been monitored in subsequently sexual generations as these plants had greatly reduced pollen viability.

Hence, a great deal of variability was observed in these *N. alata* plants regenerated from leaf explants. This somaclonal variability included variation for morphological traits (flower shape, leaf shape, plant height), pollen viability and chromosome number.

Fig. 3. Somaclonal variant for altered flower shape in the ornamental *Nicotiana alata*.

Some of the variation induced may have value for improvement of these ornamental plants. Currently, one of the most undesirable features for using ornamental types of *Nicotiana* as bedding plants is that they flower at heights of 30 cm or taller. While *N. alata* may be commercially treated with growth retardants such as daminozide, it is desirable to produce dwarf plants that will reach maturity rapidly and flower at a short height. Dwarf plants were recovered among the population of plants regenerated from leaf and petal explants. This population of dwarf-regenerated plants had concomitant decrease in pollen viability. Such reduction in pollen viability is a hindrance to integration of these dwarf somaclonal variants into a breeding program. However, for *N. alata,* we have also uncovered an alternative approach for the production of dwarfs. Some individual diploid and tetraploid regenerated plants that retained high pollen viability were used in crosses to obtain triploid progeny. Triploids flower at low plant heights compared to seed-derived controls and behaved in many respects as dwarfs. In addition to the dwarf characteristics, triploids have the added advantage of reduced seed set.

V. PLANTS FROM PROTOPLASTS AND FUSION PRODUCTS

Somaclonal variation has been documented in plants regenerated from protoplasts. This variability was first reported for the autotetraploid potato "Russet Burbank." Although only partial cytogenetic analysis has been reported, it is apparent that regenerated clones are highly variable. Over 10,000 clones have been screened and lines have been identified that are resistant to diseases to which "Russet Burbank" is susceptible [39]. Regenerates included clones that were resistant to *Alternaria solani* toxin and to *Phytophthora infestans*. These variant lines have retained the resistance phenotype through a number of vegetative generations [3]. Although it has been recognized in work on other crops that it is not necessary to use protoplasts to recover somaclonal variants, it should be recognized that all populations of plants regenerated from protoplasts probably contain somaclonal variants.

Variability is also observed in somatic hybrids produced through protoplast fusion. Interspecific or intraspecific sexual hybrids are usually genetically uniform. In most instances, particularly if hybrids are unilaterally incompatible, only certain nuclear-cytoplasmic combinations are possible. As both agronomic and nonagronomic characteristics are variables, traits important for crop improvement may be expressed in somatic hybrids and not in comparable sexual hybrids. This phenomenon emphasizes the potential value of somatic hybrids even when produced between sexually compatible species. Evans et al. [40] have compared somatic and sexual hybrids between *Nicotiana nesophila* and *N. tabacum* and reported increased phenotypic variability in the somatic hybrids.

The phenotypic variability observed in somatic hybrid plants is a reflection of genetic phenomena that occur prior to plant regeneration. Four potential sources of somaclonal variation have been identified:

1. Nuclear genetic instability in wide species combinations may result in chromosome elimination. Directional chromosome elimination has been documented in several intergeneric plant cell hybrids [41] and this instability has resulted in recovery of aneuploid plants resembling one or the other parental type for all intergeneric hybrid plants [42]. Aneuploid plants have also been recovered for more closely related interspecific somatic hybrids [43], although some of this chromosome variation may be due to the use of protoplasts isolated from long-term aneuploid cell cultures.

2. Intergenomic mitotic recombination (translocation) results in unique mixtures of genetic information. Recombination could be observed in several somatic hybrids [38] when the frequency of genetically controlled spot formation was monitored on leaves of somatic hybrids. If mitotic recombination occurs prior to shoot regeneration, this nuclear genetic variation can be detected when regenerated clones are compared. It is likely that chemical or physical agents could be used to increase the frequency of intergenomic genetic exchange.

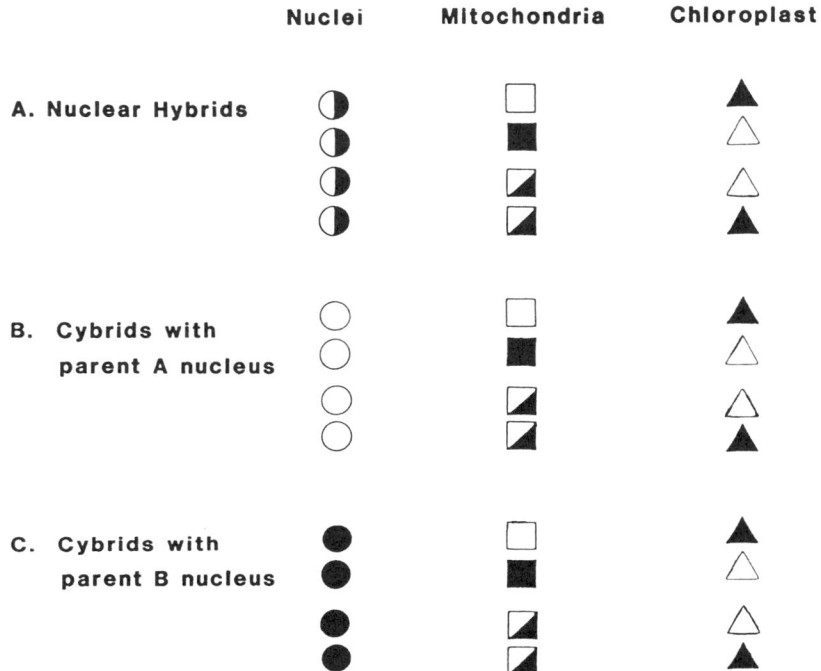

Fig. 4. Possible outcomes via somatic hybridization that are not possible via sexual hybridization. ○ and ● = nucleus; □ and ■ = mitochondria; and △ and ▲ = chloroplasts.

3. Somaclonal mutation has been observed in plants regenerated from mesophyll protoplasts [15]. This phenomenon is the result of either preexisting genetic variation in leaf cells or is induced by undefined components of the protoplast, callus, or plant regeneration media. It is as yet impossible to measure the effects of somaclonal variation on interplant variation of somatic hybrids; however, it should be noted that leaf mesophyll protoplasts, used to recover somaclonal mutations in potato, have been used to produce most somatic hybrid plants.

4. Following fusion, the organelles of the two parent lines are mixed and then recombine or segregate prior to regeneration resulting in substantial differences between clones of somatic hybrids (Fig. 4). This fourth source of variability in somatic hybrids represents one of the important differences between somatic and sexual hybrids. During fertilization, the male cyto-

plasm is preferentially excluded from the developing embryo. Thus, conventional sexual hybrids contain and express only maternally derived cytoplasmic genes. As protoplast fusion combines both parent cytoplasms into a single protoplast, fusion can be used to produce nuclear-cytoplasmic combinations not possible using conventional breeding [35].

Evidence from a number of somatic hybrids suggests that although two types of cytoplasms are initially mixed during protoplast fusion, resulting in heteroplasmons, eventually one parental type cytoplasm predominates resulting in cytoplasmic segregation. Following segregation, which may not occur until after meiosis, genetic information from this second cytoplasm is irreversibly lost. This phenomenon was first reported by examination of the fate of the large subunit of fraction 1-protein in *Nicotiana glauca* + *N. langsdorffii* somatic hybrids [44]. Random chloroplast segregation resulted in fixed langsdorffii or glauca-type fraction 1-protein in each clone prior to meiosis (10 langsdorffii-type and 7 glauca-type), except for one clone in which segregation followed self-fertilization. These two different types of cytoplasms then segregate after meiosis. Cytoplasmic segregation has been observed for other cytoplasmically controlled traits. Cytoplasmic hybrids (cybrids) containing mixed cytoplasms of cytoplasmic male sterile (cms) *Petunia hybrida* and normal *P. axillaris* segregated to male sterile and male fertile cytoplasm during the first and second meiotic cycle [45]. The male fertile character behaves as a dominant cytoplasmic trait as male fertility was expressed in putative cybrids. Similar segregation of male sterility versus male fertility has been reported in other cybrids [46]. Following analysis of four cytoplasmic characters (large subunit fraction 1-protein, cms, tentoxin sensitivity, and restriction enzyme patterns of cytoplasmic DNA), cosegregation was observed for all four traits [47]. Similarly, we have found that fraction 1-protein and tentoxin sensitivity cosegregated in *N. glauca* + *N. tabacum* somatic hybrids [48]. Cosegregation implies that following cytoplasmic segregation, the cytoplasm contains all characters from one parent or the other and, therefore, precludes widespread recombination of chloroplast genes.

Despite reports of cosegregation of cytoplasmic genes, biochemical evidence using restriction analysis of mitochondrial DNAs of some cybrids has suggested that recombination may occur between parental mitochondrial DNA prior to cytoplasmic segregation in somatic hybrids between different male sterile and fertile parent plants [49]. In these reports, mitochondrial recombination following protoplast fusion produced a wide range of morphological phenotypes controlling male sterility in the hybrid plants. Similarly, Nagy et al. [50] detected recombination in mitochondrial DNA of *N.*

tabacum + *N. knightiana* somatic hybrids resulting in novel mitochondrial DNA restriction enzyme patterns. Recombination of chloroplast DNA has recently been reported in higher plants [51]. Most likely these intriguing results of cytoplasmic segregation and recombination predict an era of extensive research on cytoplasmic genetics in higher plants. Cytoplasmic mutations for disease resistance [52] and antibiotic resistance [26] should facilitate research in this area. The production of unique nuclear-cytoplasmic combinations using protoplasts will aid in the development of novel germplasm using conventional methods.

VI. IMPLICATIONS FOR BREEDING

Based on the results presented for tomato and ornamental *Nicotiana,* an overall breeding strategy is evident for use of somaclonal variation to develop new cultivars of crops and ornamentals.

As single gene mutations and organelle gene mutations have been produced by somaclonal variation, an obvious strategy is to introduce the best available cultivars into cell culture to select for improvement of a specific character. Hence, somaclonal variation could be used to uncover new variants that retain all the favorable qualities of an existing cultivar while adding one additional trait, such as disease resistance or herbicide resistance. Work with sugar cane and tomato has already suggested that this approach is feasible. Once new R_1 variants are identified, these should be field tested in replicated plots to ascertain genetic stability. Seed should be increased at the same time to permit rapid cultivar development of promising lines. Reciprocal crosses between desirable R_1s and seed-derived controls can be used to gain an understanding of the genetic basis of the somaclonal variants. New promising breeding lines can be introduced into cell culture to add an additional character or to improve agronomic performance or ornamental appearance of a selected somaclonal variant. By using this approach, it is possible to produce new breeding lines with desirable traits in a short period of time.

VII. SUMMARY

While tissue culture cloning has found widespread application in commercial horticultural operations for micropropagation, somaclonal variation has not been so widely used. It is apparent with the unique variants that are generated, the ease of technology, and the ease of integration into a conventional breeding program, that this biotechnology tool will have increasing usefulness in horticulture and crop improvement during the next several years. The value of somaclonal variation for producing incremental changes can be equally exploited for crops or ornamentals.

REFERENCES

1. Murashige T: Plant propagation through tissue culture. Annu Rev Plant Physiol 25:135, 1974.
2. Skirvin RM, Janick J: Tissue culture-induced variation in scented *Pelargonium* spp. J Am Soc Hort Sci 101:281, 1976.
3. Shepard JF, Bidney D, Shahin E: Potato protoplast in crop improvement. Science 208:17, 1980.
4. Larkin PJ, Scowcroft WR: Somaclonal variation and eyespot toxin tolerance in sugarcane. Plant Cell Tissue Organ Culture 2:111, 1981.
5. Evans DA, Sharp WR: Single gene mutations in tomato plants regenerated from tissue culture. Science 221:949, 1983.
6. Evans DA, Sharp WR, Medina-Filho HP: Somaclonal and gametoclonal variation. Am J Bot 71:759, 1984.
7. Chaleff RS: "Genetics of Higher Plants." Cambridge: Cambridge University Press, 1981.
8. Orton TJ: Spontaneous electrophoretic and chromosomal variability in callus cultures and regenerated plants of celery. Theor Appl Genet 67:17, 1983.
9. Ben-Jaacov J, Langhans RW: Raplid multiplication of *Chrysanthemum* plants by stem proliferation. Hort sci 7:289, 1972.
10. Vajrrabhaya T: Variations in clonal propagation. In Arditti J (ed): "Orchid Biology: Review and Perspectives": New York: Columbia University Press, 1977, p 179.
11. Engler DE, Grogan RG: Variation in lettuce plants regenerated from protoplasts. J Hered 75:426, 1984.
12. Pfeiffer JW, Bingham ET: Comparisons of alfalfa somaclonal and sexual derivatives from the same genetic source. Theor Appl Genet 67:263, 1984.
13. Evans DA, Bravo JE: Phenotypic and genotypic stability of tissue cultured plants. In Zimmerman RH, Griesbach RJ, Hammerschlag FA, Lawson RH (eds): "Tissue Culture as a Plant Production System for Horticultural Crops." Netherlands: Martinus Nijhoff Publishers, 1986, p 73.
14. Bhankaran S, Smith RH, Finer JJ: Ribulose bisphosphate carboxylase activity in anther-derived plants by *Saintpaulia ionantha* Wendl Shag. Plant Physiol 73:639, 1983.
15. Shepard JF: The regeneration of potato plants from leaf cell protoplasts. Sci Am 246:154, 1982.
16. Brook Houzen PD, Cook JP, Cowley CR, Hollingsworth MD: Analysis of corn mutants derived from *in vitro* culture. In "Agronomy Abstracts." Las Vegas, Nevada: Am Soc Agron, 1984, p 60.
17. Smith SM, Street HE: The decline of embryogenic potential as callus and suspension cultures of carrot (*Daucus carota* L.) are serially subcultured. Ann Bot 38:233, 1974.
18. Singh BD, Harvey BL, Kao KN, Miller RA: Karyotypic changes and selection pressure in *Haplopappus gracilis* suspension cultures. Can J Genet Cytol 17:109, 1975.
19. Skirvin RM: Natural and induced variation in tissue culture. Euphytica 27:241, 1978.
20. D'Amato F: Cytogenetics of differentiation in tissue and cell cultures. In Reinert J, Bajaj YPS (eds): "Applied and Fundamental Aspects of Plant Cell, Tissue, and Organ Culture." Berlin: Springer Verlag, 1977, p 343.

21. Bennici A, Buiatti M, D'Amato F, Pagliai M: Nuclear behavior in *Haplopappus gracillus* callus grown *in vitro* on different culture media. Coloq Int CNRS 193:245, 1971.
22. Brossard D: Influence of kinetin on formation and ploidy level of buds arising from *Nicotiana tabacum* pith tissue grown *in vitro*. Z Pflanzenphysiol 78:323, 1976.
23. Reisch B, Bingham ET: Plants from ethionine-resistant alfalfa tissue cultures: Variation in growth and morphological characteristics. Crop Sci 21:783, 1981.
24. Sacristan MD, Melchers G: The caryological analysis of plants regenerated from tumorous and other callus cultures of tobacco. Mol Gen Genet 105:317, 1969.
25. Heinz DJ, Mee WP: Morphologic, cytogenetic, and enzymatic variation in *Saccharum* species hybrid clones derived from callus tissue. Am J Bot 58:257, 1971.
26. Maliga P: The need and the search for genetic markers in plant cell cultures. In Sala F, Parisi B, Cella R, Ciferri O (eds): "Plant Cell Cultures: Results and Perspectives:" Amsterdam: Elsevier/North Holland, 1980, p 170.
27. Evans DA, Sharp WR: Somaclonal variation in agriculture. Bio/Technology 4:528, 1986.
28. Sibi M: Multiplication conforms, non-conforms. Le Selectionneur Francais 26:9, 1978.
29. Gengenbach BG, Connelly JA, Pring DR, Conde MF: Mitochondrial DNA variation in maize plants regenerated during tissue culture selection. Theor Appl Genet 59:161, 1981.
30. Krishnamurthi M, Tlaskal J: Fiji disease resistant *Saccharum officinarum* var. Pindar subclones from tissue cultures. Proc Int Soc Sugar Cane Technol 15:130, 1974.
31. Matern U, Strobel G, Shepard JF: Reaction to phytotoxins in a potato population derived from mesophyll protoplasts. Proc Natl Acad Sci USA 75:4935, 1978.
32. Karp A, Nelson RS, Thomas E, Bright SWJ: Chromosome variation in protoplast-derived potato plants. Theor Appl Genet 63:265, 1982.
33. Sacristan MD: Resistance response to *Phoma lingam* of plants regenerated from selected cell embryogenic cultures of haploid *Brassica napus*. Theor Appl Genet 61:193, 1982.
34. Padmanabhan V, Paddock EF, Sharp WR: Plantlet formation from *Lycopersicon esculentum* leaf callus. Can J Bot 52:1429, 1972.
35. Gleba YY: Nonchomosomal inheritance in higher plants as studied by somatic hybridization. In Sharp WR, Larsen PO, Paddock EF, Raghaven V (eds): "Plant Cell and Tissue Culture Principles and Applications." Columbus: Ohio State University Press, 1979, p 775.
36. Miller SA, Williams GR, Medina-Filho HP, Evans DA: A somaclonal variant of tomato resistant to Race 2 of *Fusarium oxysporum* f. sp. *lycopersici*. Phytopathology 75:1354, 1985.
37. Bendich AJ, Gauriloff LP: Morphometric analysis of cucurbit mitochondia. The relationship between chondriome volume and DNA content. Protoplasma 119:1, 1984.
38. Evans DA, Wetter LR, Gamborg OL: Somatic hybrid plants of *Nicotiana glauca* and *Nicotiana tabacum* obtained by protoplast fusion. Physiol Plant 48:225, 1980.
39. Shepard JF: Mutant selection and plant regeneration from potato mysophyll protoplast. In Rubenstein, Gengenbach B, Phillips RL, Green CE (eds): "Genetic Improvement of Crops." University of Minnesota Press, Minnesota, 1980.
40. Evans DA, Flick CE, Kut SA, Reed SM: Comparison of *Nicotiana tabacum* and *Nicotiana nesophila* hybrids produced by ovule culture and protoplast fusion. Theor Appl Genet 62:193, 1982.
41. Kao KN: Chromosomal behavior in somatic hybrids of soybean-*Nicotiana glauca*. Molec Gen Genet 150:225, 1977.
42. Hoffmann F, Adachi T: "Arabidobrassica." Chromosomal recombination and morphogenesis in asymmetric intergeneric hybrid cells. Planta 153:586, 1981.

43. Bravo JE, Evans DA: Protoplast fusion for crop improvement. Plant Breed Rev 3:193, 1985.
44. Chen K, Wildman SG, Smith HH: Chloroplast DNA distribution in parasexual hybrids as shown by polypeptide composition of fraction-1 protein. Proc Natl Acad Sci USA 74:5109, 1977.
45. Izhar S, Tabib Y: Somatic hybridization in *Petunia*. Part 2: Heteroplasmic state in somatic hybrids followed by cytoplasmic segregation into male sterile and female fertile lines. Theor Appl Genet 57:241, 1980.
46. Gleba YY, Evans DA: Genetic analysis of somatic hybrid plants. In Evans DA, Sharp WR, Ammirato PV, Yamada Y (eds): "Handbook of Plant Cell Culture." New York: Macmillan, 1984, p 322.
47. Aviv D, Fluhr R, Edelman M, Galun E: Progeny analysis of the interspecific somatic hybrids: *Nicotiana tabacum* (cms) + *Nicotiana sylvestris* with respect to nuclear and chloroplast markers. Theor Appl Genet 56:145, 1980.
48. Flick CE, Evans DA: Evaluation of cytoplasmic segregation in somatic hybrids in the genus *Nicotiana:* Tentoxin sensitivity. J Hered 73:264, 1982.
49. Belliard G, Vedel F, Pelletier G: Mitochondrial recombination in cytoplasmic hybrids of *Nicotiana tabacum* by protoplast fusion. Nature 281:401, 1979.
50. Nagy F, Torok I, Maliga P: Extensive rearrangements in the mitochondrial DNA in somatic hybrids of *Nicotiana tabacum* and *Nicotiana knightiana*. Mol Gen Genet 183:437, 1981.
51. Medgyesy P, Fejes E, Maliga P: Interspecific chloroplast recombination in a *Nicotiana* somatic hybrid. Proc Natl Acad Sci USA 82:6960, 1985.
52. Gengenbach BG, Green CE, Donovan CM: Inheritance of selected pathotoxin resistance in maize plants regenerated from cell cultures. Proc Natl Acad Sci USA 74:5113, 1977.

Biotechnology in Agriculture, pages 225–248
© 1988 Alan R. Liss, Inc.

Encapsulated Plant Embryos

Keith Redenbaugh, Jo Ann Fujii, and David Slade

Plant Genetics, Inc., Davis, California 95616

———◆◆———

———◆◆———

I. ARTIFICIAL SEEDS

A. Potential of Artificial Seeds

1. Applications. The major reason for developing artificial seeds is to achieve a more desirable stand establishment and yield. Techniques such as vegetative propagation and hybrid seed production that significantly increase uniformity in a variety are not available or are too expensive for a great many crops. Artificial seeds are a potential delivery system to provide an alternative to current high-cost vegetative propagation and to allow for planting of hybrid "seed." Based on estimates for materials and labor for embryo production and encapsulation as well as on propagation rates of somatic embryos, artificial seeds should be available at low cost and high volume while remaining competitive with the cost structure of seeds and maintaining the genetic uniformity of vegetative propagation. Cost of an artificial seed has been estimated to be 0.026¢ per unit [1].

In theory, the ability to use a small (several millimeter) propagule rather than a large plantlet (tens of centimeters) provides tremendous flexibility for plant propagation. Choice of species for clonal multiplication is broadened and crops can be engineered more specifically for field and environmental conditions if a seed-size propagule is available. The size of the propagule is crucial, as well, for storage, handling, shipping, and planting, particularly when very large numbers are required. The larger the plant material the more difficult and costly these steps become. Direct sowing in the field will also be possible with a seed-size propagule, which would eliminate any acclimation step normally required for transplants.

Artificial seeds will be useful for propagation of hand-pollinated hybrids, elite germplasm, and genetically engineered plants, particularly those with sterile or unstable genotypes The use of artificial seeds potentially eliminates the requirement for meiotically stable integration of the genetic changes and may even be advantageous for product protection. Artificial seeds will also be useful in cases where seed fertility is reduced, such as with wide hybridization designed to incorporate diverse genetic material in a unique genotype. Wide crosses may be produced using sexual or asexual means (protoplast fusion, cybrid production, or single-gene transfers).

2. Crops. The initial use of artificial seeds will be with medium-to-high-value crops, such as flower and ornamental species, vegetables, and for use in breeding programs such as alfalfa and forest trees (Table I). Researchers are beginning to identify potential crops for focus, such as Kamada [2], who suggested the following crops for artificial seed use, basing his choices on economic considerations: "vegetative propagation plants or perennial plants such as potato, sweet potato, fruit trees and woody plants; dioecious plants

TABLE I. Potential Crops for Artificial Seeds

Category 1—Strong technological basis
Good somatic embryogeny system currently available

Alfalfa	Oil palm
Caraway	Orchardgrass
Carrot	Orange
Celery	*Panicum*
Coffee	*Pennisetium*
Eggplant	Walnut

Category 2—Strong commercial basis
Economic importance based on value of vegetative propagation

Asparagus	Impatiens
Begonia	Lettuce
Broccoli	Loblolly pine
Cauliflower	Petunia
Corn	Potato
Cotton	Rice
Cucumber	Soybean
Cyclamen	Spinach
Douglas fir	Sugar cane
Garlic	Tobacco
Geranium	Tomato
Gerbera daisy	Watermelon
Grape	

where either sex alone is useful from an agricultural viewpoint, such as asparagus and kiwi fruit; and tetraploid plants which can be hardly bred, such as potato and alfalfa."

Currently, there are relatively few species for which predictable embryo regeneration and plant production are achievable (Table I, Category 1). These species are characterized by the intensive amount of research that has been done to improve the quality of the embryos. Even for these crops, however, the quality of the embryos has not been sufficient for creating commercial artificial seed products. A second list of crops are those for which vegetative propagation is desirable, but for which good embryogeny systems have not yet been developed (Table I, Category 2). This list can serve as a focus for directing artificial seed research.

For example, the production of new lettuce varieties generally begins with an F_1 cross followed by 8 to 12 generations of selfing (to insure homozygosity) and evaluations prior to varietal release. With lettuce artificial seeds, single plant selections could be made from selected F_1 plants. Evaluation and trialing could begin immediately and no inbreeding would be needed to fix desirable genes.

A second example is alfalfa, for which maximizing heterozygosity, rather than homozygosity, is the objective. Traditionally, alfalfa varieties are produced by selecting seed sources and/or specific genotypes and crossing these to produce breeder's seed. Heterozygosity is greatest at this seed generation. Two to four more years are required to evaluate and produce enough seed for sale. During each subsequent generation, the heterozygosity decreases, particularly if the selection index was high. The final seed sold is of lower quality than the breeder's seed. Large-scale clonal propagation using artificial seeds would significantly reduce the number of generations needed for commercial seed production and increase the level of heterozygosity.

3. Advantages of the capsule. One of the primary functions of the capsule is to protect the somatic embryo during storage, shipping, handling, and planting. Even though naked somatic embryos can be planted directly in soilless mix in the greenhouse with subsequent plant production (Fujii, Slade, and Redenbaugh, unpublished results), embryos would be damaged during large-scale, mechanical operations without a protective coating. The capsule itself may be an inert component of the artificial seed or it may have biological activity, such as a fertilizer which complexes the capsule gel or a carbohydrate source used for dry-coating. The capsule also has the potential to hold and deliver beneficial adjuvants such as growth-promoting bacteria, nematodes, plant nutrients, growth control agents, fertilizers and pesticides, all of which could be placed precisely in the individual plant's rhizosphere.

B. Definition of Artificial Seeds

An artificial seed is a novel analog to botanic seed consisting of a somatic embryo surrounded by an artificial seed coat. The seed coat must be nondamaging to the embryos, protect the embryos from mechanical damage during handling, and allow germination and conversion (embryo-to-plant development) to occur, without introducing variation-inducing components. The artificial seed may contain and deliver nutrients in the form of an artificial endosperm as well as developmental control agents and other components necessary for germination and conversion. An outer or second seed coat may be required to slow capsule desiccation. For many crops such as vegetables, it will be necessary that single-embryo artificial seeds be produced. The precise constituents of an artificial seed will vary from crop to crop.

Artificial seeds may conceivably be hydrated or desiccated in final form. Progress has been greatest to date with hydrated artificial seeds, mainly because the hydrated capsule essentially recreates a small nutrient-containing gel environment. Although there has just now been success in desiccating somatic embryos and recovering plants, the first applications of artificial seed technology most likely will be a hydrated artificial seed used in a breeding program (such as alfalfa) or in a greenhouse as a transplant vegetable (such as celery).

However, as greater understanding of embryogeny is obtained and more mature and complete somatic embryos are formed, then greater success is expected with embryo desiccation. Conceptually, a desiccated artificial seed would be more analogous to true seeds and have a longer shelf life. Additionally, the desiccation process may contribute to embryo maturation and therefore aid embryo survival and conversion. The developmental stage of even the most highly matured somatic embryo, to date, may be equal only to that of a zygotic embryo that has reached the stage where it is capable of precocious germination. Such zygotic embryos are not fully developed and lack the vigor of mature embryos. Gray et al. [3] compared the difference between somatic and zygotic embryos: "the latter typically cease growth, becoming quiescent or dormant as water is lost, storage tissues mature, and the seed coat hardens. This arrested growth phase is the major factor accounting for the efficient storage and handling qualities of seed." Ultimately, the use of hydrated or desiccated artificial seeds will depend on the species and growing situation.

C. Fundamental Problem

The fundamental problem for artificial seeds is to produce a population of somatic embryos that germinates and forms seedling-quality plants. Without somatic embryos of vigor and thriftiness equal to zygotic embryos, the ultimate objectives of obtaining stand establishment and improved yield will not be achieved. For commercial applications, artificial seeds need to have germination and plant development frequencies and rates comparable to true seeds. Research reports on somatic embryogeny and artificial seeds need to address this fundamental problem and clearly discuss results in appropriate, well-defined terms. In a recent article, Gray et al. [3] provided clear, specific data on two degrees of germination: that without further plant growth and that with production of viable plants. Other researchers using the term germination have not defined it or have not clearly stated whether complete plants were formed [2,4–7].

II. HISTORICAL DEVELOPMENT AND CURRENT PROGRESS OF ARTIFICIAL SEEDS

Murashige discussed the concept of artificial seeds during talks and discussions in the early 1970s (Murashige, personal communication) and formally presented his ideas for the first time at the *Symposium on Tissue Culture for Horticultural Purposes* in Ghent, Belgium, September 6–9, 1977 [8]. Other researchers working in the mid-1970s on somatic embryogenesis for crop propagation were Walker, then at Monsanto Company (personal communication), and Lawrence at Union Carbide. Fluid drilling technology was the major emphasis of Lawrence's efforts for delivering somatic embryos [9]. However, these three researchers did not report any experimental data on artificial seed technology.

Gray organized the first symposium on artificial seeds for the 22nd International Horticultural Congress in August 1986 in Davis, California. His objective was to bring together the critical issues of artificial seed technology and "collectively . . . present a coherent view of the field" [10]. Topics covered were zygotic embryogeny and control of embryo maturation, cost estimates for spin-bioreactor systems (unpublished), bioreactor production of alfalfa somatic embryos, encapsulation of somatic embryos, and somatic embryo desiccation.

A. Fluid Drilling

Research with a focus on fluid drilling of somatic embryos was reported by Drew [11]. His objective was "to develop a method of bulk handling many small plantlets, perhaps millions, and avoiding the need for individual handling." Drew was able to produce three plants from carrot embryos placed on filter paper bridges suspended over carbohydrate-free, nutrient medium. In a second experiment, three sucrose-loaded embryos produced plants on agar-based, sucrose-free White's medium [12] in tightly sealed, clear plastic boxes. Drew found that "perhaps the most crucial one [problem] is the very slow rate of development of plantlets derived from culture." He suggested that if the quality of the embryos could be improved to make them photosynthetically competent more quickly, then fluid drilling could be used for field delivery of carrot somatic embryos. However, he did not report using fluid drilling for planting carrot embryos.

Baker [13] presented a thorough review of a fluid drilling approach for delivery of somatic embryos. She was able to plant carrot somatic embryos mixed in a sucrose/hormone/fluid gel directly in the greenhouse. Four percent of the embryos survived for 7 days, but further development and plant conversion were not possible because of embryo desiccation and death. Without sucrose, the embryos did not even survive in the gel and no emergence occurred. Under- and overwatering appeared to be a major problem for carrot embryo survival.

B. Desiccated Coatings

Kitto and Janick first reported on coating embryogenic suspensions of carrot in polyoxyethylene wafers at the 79th Annual Meeting of the American Society for Horticultural Science [14]. The embryogenic suspensions contained "cells, cell aggregates, callus clumps and embryos of varying maturity." Of much interest were their reports that 3% of the coated somatic embryos survived desiccation. Without the coating the survival was "nil." Although the term "survival" was not defined, the authors reported (without data) that the "carrot embryos germinated and grew after encapsulation and rehydration" [5]. In a second report, Kitto and Janick [6] improved embryo survival with high sucrose, high inoculum density, or chilling treatments with or without abscisic acid. Again, survival was not defined. However, they reported that the "carrot embryos grew after rehydration" and showed a petri dish containing leaves from rehydrated embryos. Gray et al. [3] suggested that two plants were produced from 20 wafers in their experiments. Because single embryos were not coated, it was unclear as to whether the surviving embryos existed prior to coating or if the embryos formed de novo from surviving callus after rehydration. An earlier report on desiccation, rehydration, and survival of carrot callus/embryos was also unclear on this [15]. One likely possibility is that very small embryos survive the desiccation and, upon rehydration, produce secondary embryos as well as embryogenic callus.

Recently, Gray et al. [3] reported desiccating orchardgrass (*Dactylis glomerata* L.) somatic embryos to 13% water content. After 21 days storage at 23° C, they achieved 4% germination and conversion. The conversion frequencies of the desiccated embryos decreased from an initial 32% without storage, to 8% after 7 days storage, then 4% after 21 days. Desiccation was done in empty petri dishes at 23° C in the dark at a relative humidity of 70% ± 5%. Gray et al. wrote "during desiccation, somatic embryos decreased in size, became yellowish and brittle, and their outer cell walls were collapsed." Because the embryos were desiccated, the authors considered them to be quiescent. They found that white, opaque, well-developed embryos between 1 and 1.5 mm in length produced plants. Smaller or larger embryos failed even to germinate. Despite the low conversion frequencies and the short storage time for the embryos, their report was very encouraging suggesting that, with the proper components, the embryos could be dry-coated to produce a desiccated artificial seed.

C. Hydrogels

Redenbaugh et al. [16,17] discovered that hydrogels such as sodium alginate could be used to produce single-embryo artificial seeds. In vitro conversion frequencies as high as 86% were reported for alfalfa artificial seeds. Artificial seeds were also planted in the greenhouse with conversion up to 20% [1,17,34]. This research is described further in the following sections.

Yamakawa discussed artificial seeds at a conference in Japan in 1983 saying, "it is epoch-making if a pseudo-embryo prepared from each cultured cell could be made storage-stable by some treatments and marketed and a farmer could readily regenerate this pseudo-embryo into a plant body, i.e., pseudo-embryos obtained by asexual propagation could substitute for conventional seeds. Then the present breeding system might be entirely changed" [18]. Subsequently, he discovered ongoing research in this area at Plant Genetics, Inc., in the United States [19] and began to emphasize the importance of such technology in Japan. He viewed the limitation for this technology to be the tremendous difficulty "to transport a meristem formed in vitro to a farm for transplantation."

In a news article in Nature on artificial seeds, Walgate [20] quoted Demarly (Orsay, France), "we can already coat somatic embryos of lucerne [a legume] in a kind of jelly, and get 100 per cent germination." Greater details were not given. Sylvia Bianchi (personal communication) indicated that selected lucerne (alfalfa) somatic embryos had been encapsulated in sodium alginate. Somatic embryos from other crops were also encapsulated, but the results were not as good as those with lucerne.

Lutz et al. [7] reported encapsulating carrot somatic embryos in a gel matrix, presumably a hydrogel. The embryos germinated and grew out of the capsule; however, they were not specific as to whether plants were produced. Their conclusion was "embryo quality is a limiting factor in developing a delivery system."

In a Japanese patent application, Hama [4] described research at Lion Company in Japan in which artificial seeds were defined as consisting of an adventitious embryo with nutrients, an antibacterial solution, and a water-absorptive polymeric material encased in a hydrogel. Without the antibacterial agent or the water-absorptive polymeric material, no embryo germination occurred when the artificial seeds were sown on vermiculite. The antibacterial solution used was copper hydroxyquinoline with penicillin, streptomycin, or zineb with chloramphenicol. The polymeric material was either polyacrylate or saponified starch/acrylonitrile graft copolymer. Capsules were produced with sodium alginate dropped into calcium chloride or with kappa-carrageenan dropped into potassium chloride, using the system described by Redenbaugh [21]. Hama stated "the artificial seeds . . . broke the coating within approximately 10 days and showed normal growth thereafter." However, "normal growth" was not defined and it was not clear whether whole plants were produced. Later, Nishimura [22] reported that Lion Co. produced plants from artificial seeds planted on vermiculite under non-sterile, high humidity conditions. Unfortunately, further details were not presented.

Kamada [2] provided this definition—"an artificial seed comprises a capsule prepared by coating a cultured matter, a tissue piece or an organ which can grow into a plant body and nutrients with an artificial film." His artificial seed concept consisted of "an external film for strengthening the seed . . . an internal film for encapsulating (a gel such as calcium alginate) . . . nutrients required for the growth of the cultured matter and plant hormones for controlling germination, i.e., corresponding to the albumen of a conventional seed; and . . . a callus, an adventitious bud or an adventitious embryo which can grow into a plant body." Much of Kamada's description of the encapsulation system was based on Redenbaugh [21]. Using the alginate system, he described a dual-nozzle encapsulation device in which a solution containing embryos, nutrients, and hormones passed down an inner tube while sodium alginate moved along an outer tube. The two solutions met at the tip of the apparatus and drops were formed. Approximately half of the capsules formed contained somatic embryos. These artificial seeds were sown on vermiculite under high humidity and asceptic conditions; however, the constituents of the capsule (nutrients, carbohydrates, etc.) were not defined. Germination frequencies of 3–10% were achieved for carrot, but the report was unclear as to whether whole plants were obtained.

Gupta and Durzan [23] used sodium alginate to encapsulate loblolly pine somatic embryos. The embryos were stored at 4°C for 4 mo in the dark. Upon return to the light at 20°C, the embryos produced chlorophyll indicating survival. However, conversion was not obtained. Kirin Brewery Co., Ltd., in Japan has been working on developing artificial seed technology for celery, lettuce, rice, and other crops using a hydrated, hydrogel encapsulation system (M. Sanada, personal communication).

Upadhyaya et al. [24] modified seed hoppers of a belt-type planter (Stanhay) and a vacuum planter (Gaspardo) to handle and plant hydrogel capsules. Capsules without a hydrophobic coating (Elvax) did not flow uniformly and tended to stick together regardless of modifications made. Elvax-coated alginate capsules could be made to flow evenly when modifications were made to the seed hoppers. For the Stanhay, a wire wiggler and a wire frame were designed to agitate the capsules in the seedbox. The wire frame was more effective in planting the capsules, leaving only 16% of the capsules in the seedbox (compared to 40% for no agitation and 33% for the wire wiggler). An air nozzle was added to the Stanhay to blow off capsules sticking to the belt. For the Gaspardo, a double-taper, open, helical agitator was designed to provide a more even flow of capsules. Only 15% of the capsules were left in the seedbox using this modification. Upadhyaya et al. [24] concluded their report with "both vacuum seeders like Gaspardo and belt seeders like Stanhay can easily be retrofitted to adequately meter gel encapsulated propagules." They did suggest, however, that "a seeder specifically designed to accurately meter the gel encapsulated propagules would be highly desirable."

D. Other Techniques

Seed-coating technology (pelletization), which has gained general acceptance for a number of seed crops over the past decade, has been proposed for coating somatic embryos. Durzan [25] suggested that for forest tree species, "embryos may be pelleted by special coatings to form artificial seeds that can be easily handled for mass planting." He suggested that nursery plantings might require fungicides to be placed in the pellet containing the embryo. Pelletization would require that the somatic embryos be desiccated since the process uses a dry coating consisting of diatomaceous earth, binders and other inert compounds. Pellatization may be practical for sizing artificial seeds that have been desiccated and covered with a protective coating, such as polyoxyethylene. However, no reports on pellatization of somatic embryos have appeared.

Another proposed delivery method for somatic embryos is a seed tape [26]. Somatic embryos could be precision placed in a tape under sterile conditions that would facilitate storage. Embryos could be hardened in the tape during storage. The "embryo" tapes would then be unrolled in greenhouses or fields, depending on the growing parameters of the crop and the robustness of the embryos.

Finally, new technology has been used to coat primed or pregerminated seeds and this method may have application to either hydrated or partially desiccated embryos. Royal Sluis Company has introduced a product, Quick Pill, which uses a trade secret process to coat pregerminated celery seeds in a semihydrated coating [27]. Quick Pill has a short shelf life of only 2 wk but achieves 95% stand establishment in the greenhouse. No reports have appeared using this process on somatic embryos.

E. Patents

Three US patents have been issued on artificial seeds. The first two by Redenbaugh [21,28] covered hydrated artificial seeds and various adjuvants in the capsules. The focus of the two patents was on the use of a variety of hydrogels for encapsulation of somatic embryos of several crops. The third by Janick and Kitto [29] was on desiccated artificial seeds, which included processes to harden, coat, and desiccate carrot somatic embryos.

F. Directions for Artificial Seed Research

Research on artificial seeds has been limited to two systems of encapsulation/coating: polyoxyethylene [5] and hydrated hydrogels [17]. Certainly, other methods for producing an artificial seed coat will be discovered, but currently researchers have focused on the alginate system for alfalfa [1,20,30–32], carrot [2,4,7], celery [17], and loblolly pine [23] with various responses from the embryos. Others such as Ammirato [33] and Gray et al. [3] have concentrated on somatic embryogeny and not reported on encapsulation. In reviewing reports on artificial seeds and somatic embryogeny, the common conclusion was that the quality of the embryo was the limiting factor for application. Additionally, when embryos were encapsulated, the interaction between somatic embryos and the encapsulation system introduced additional parameters. Nonvigorous embryos, such as celery, that had high in vitro conversion frequencies (50–80%) showed reduced frequencies when encapsulated, whereas vigorous embryos, such as alfalfa, showed no such reduction (Fujii, Slade, and Redenbaugh, unpublished data). As progress is made in somatic embryogeny, the encapsulation methodology must be adapted to fit the requirements of the embryos. Dual progress in somatic embryogeny and encapsulation is required for the creation of commercial artificial seeds. The remainder of this paper will focus on this concept and the artificial seed technology developed by the authors.

III. METHODS FOR HYDROGEL ENCAPSULATION OF PLANT EMBRYOS

A. Capsule System

Initial research focused on developing methods to encapsulate alfalfa somatic embryos. Four encapsulation methods were tested: two types of gelation, complex coacervation, and interfacial polymerization. Alfalfa somatic embryos survived only when encapsulated using one of the two gelation methods: gel complexing via a dropping procedure or molding via reduction in temperature. Twenty-seven water soluble hydrogel/gel combinations were then tested for capsule production and embryo survival. Six gels were found to be nondamaging to the embryos. The term ''survival'' was used initially to describe alfalfa embryos that remained green and continued to enlarge (without radicle elongation) after encapsulation. During initial encapsulation research, the quality of alfalfa embryos was very poor and only low in vitro conversion frequencies (<1%) could be obtained. Conversion to

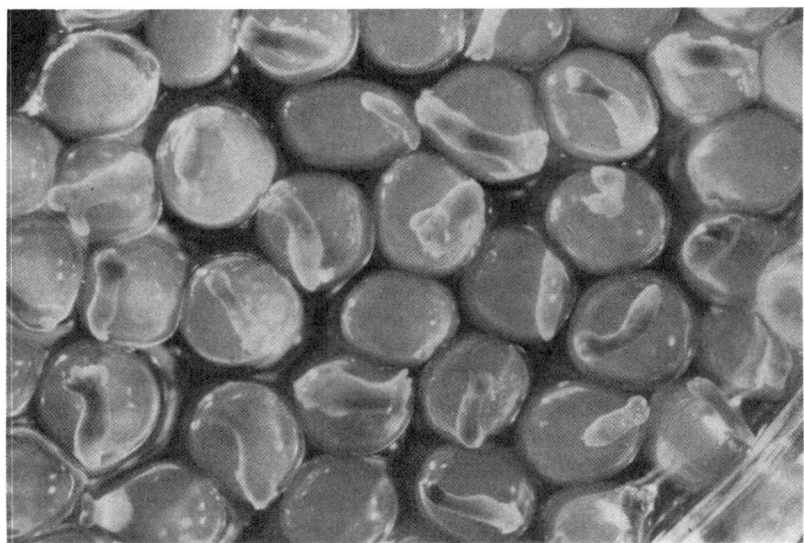

Fig. 1. Alfalfa somatic embryos encapsulated in calcium alginate. Magnification ×2.

plants did occur when alfalfa artificial seeds were encapsulated with alginate, alginate with gelatin, or carrageenan with locust bean gum. The embryos were not hardy enough to survive encapsulation using other gels. Sodium alginate was chosen as the major encapsulation gel, because of its gelation properties, ease of use, nontoxicity, and low cost [30].

Alfalfa somatic embryos were mixed with sodium alginate (2% w/v) and dropped into a calcium nitrate solution (100 mM). Surface complexation began immediately and the drops were gelled completely in 30 min (Fig. 1). Alternatively, the embryos could be mixed in a temperature-dependent gel such as Gel-rite™, placed in the well of a microtiter plate, and gelled as the temperature was lowered [1]. Initially, conversion frequencies were very low even with selected alfalfa embryos (less than 1%). As improvements were made in somatic embryogeny (see next section), the in vitro frequencies increased significantly.

Sodium alginate complexes in the presence of di- and trivalent metal cations to form calcium alginate. Metal cations form ionic bonds between carboxylic acid groups on the guluronic acid molecules of the alginate. Alginates that are rich in guluronic acid form harder capsules than alginates composed of significant levels of mannuronic acid. Sodium alginate will go into solution and remain stable at room temperature. It does not require heat to produce a gel, but begins complexing immediately when brought into contact with metal cations. The hardness of the capsule is a function of the guluronic:mannuronic acid ratio, the cation, and the complexing time. The size of the capsule can be controlled by the viscosity of the sodium alginate and by the inside diameter of the nozzle used to form the drops [31].

B. Artificial Endosperm

Another potential component of artificial seeds is a synthetic endosperm. For albuminous species, such as carrot or celery, an artificial endosperm will likely be required to provide nutrients and carbon sources necessary for optimum germination and conversion. Drew [11] found that sucrose was a requirement for carrot somatic embryo germination and plant development. Exalbuminous crops, such as alfalfa, may need no endosperm. Initial experiments testing conversion frequencies for both alfalfa and celery artificial seeds in an in vivo environment (sand trays or peat/vermiculite plugs in a greenhouse) resulted in low conversion frequencies (7% and 10%, respectively) as compared with much higher in vitro frequencies. Artificial seeds for subsequent in vivo conversion experiments were planted in either a greenhouse mix or vermiculite in aluminum trays and grown nonsterilely at 25°C in an incubator or in a greenhouse [1,17].

Alfalfa was chosen for development of an artificial endosperm despite its being an exalbuminous species. Although the conversion frequency of alfalfa is low under in vivo conditions, its somatic embryogenesis system is fairly well characterized. Two approaches were tested: to add complex carbohydrates to the gel capsule to minimize leaching and to microencapsulate nutrients to provide controlled release inside the artificial seed. Various starches, crude alginates, and sucrose microcapsules (produced via two methods, solvent evaporation or complex coacervation) were used as additives. Unrefined, crude alginate (Protatek EFN, Protan Company, Norway) used to form the capsules was the only additive that resulted in an increase in in vivo conversion. The other components were ineffective [30].

C. Hydrophobic Coatings

Calcium alginate capsules were very sticky and dried rapidly when exposed to air. This caused difficulty with handling and machine planting [24]. Seven coating procedures were tested on alginate capsules to prevent capsule desiccation. One compound, Elvax™ 4260 (Dupont, developed by Dr. Zoila Reyes, SRI International) was very hydrophobic and showed a significant impediment to capsule drying, such that half the capsule weight could be retained after 4 days in an unsealed container [30,34]. Furthermore, the degree of tackiness was reduced such that coated capsules could be

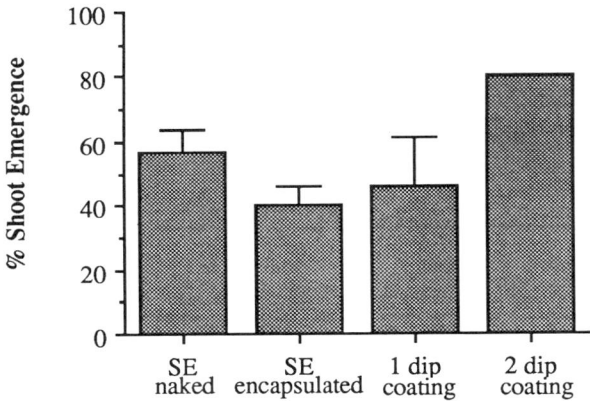

Fig. 2. Shoot emergence from alfalfa somatic embryos from alginate capsules coated with Elvax™ 4260 polymer. (Reprinted from HortScience [30].)

planted using a Stanhay seed planter [17,24]. Encapsulated alfalfa somatic embryos were coated with Elvax without reducing shoot emergence (Fig. 2).

IV. SOMATIC EMBRYOGENY

Once a basic encapsulation system was designed, then modifications could be made to improve the capsule structure, depending on the requirements of embryos for conversion. Because conversion frequencies for alfalfa were so low initially, it was difficult to determine the effect of such modifications. Consequently, the focus for artificial seed development was directed toward improving embryo quality. The encapsulation system was repeatably tested as improvements were made in the quality of the embryos.

In theory, somatic embryogeny spans the process beginning with the onset of embryogenesis and continuing to the production of fully mature, developmentally arrested embryos. The use of the term ''somatic embryogeny'' is important in order to emphasis the necessity of producing high-quality embryos. Most reports in the literature have focused on somatic embryogenesis and the production of a population of embryos, some of which will continue development and produce plants, but most of which are abnormal, incomplete embryos. Without a selection pressure such as plant conversion, protocols developed for embryogenesis may be limited to initial embryo formation and the embryos produced, despite normal morphology, may develop into plants with severe abnormalities.

A. Alfalfa Somatic Embryogeny

For alfalfa somatic embryogenesis, callus was initiated from petioles placed on Schenk and Hildebrandt (SH) medium [35] containing 25 μM naphthaleneacetic acid, 10 μM kinetin, and 8 g/liter agar. Callus was induced to form embryos on SH medium containing 50 μM 2,4-dichlorophenoxyacetic acid and 5 μM kinetin. When placed on SH medium with 10 mM ammonium and 30 mM proline for 21 days, 1–5-mm embryos formed [36]. Somatic embryos were placed on half-strength SH medium with 1.5% maltose (instead of sucrose) and scored for conversion after 28 days. Maltose was used in the conversion medium after it was discovered to be a potent substitute for proline during regeneration [37].

The term "conversion" was coined to clarify development of the embryos. "Germination" of somatic embryos was nebulous since it could refer to anything from simple radicle elongation (without further development) to complete plant formation. As discussed earlier, the growth response of somatic embryos is critical for propagation. Results need to be described in precise terms. For crops with good somatic embryogeny systems, such as alfalfa and celery, the in vitro conversion frequencies were high. Embryos were characterized by a bipolar structure with root and shoot axis, well-developed cotyledons, and lack of callusing or swelling. Upon germination, rapid root production with secondary and tertiary branching was observed, the shoot meristem began to grow, cotyledons expanded, and the first true leaves appeared (conversion). Embryo conversion frequency was the percent of the somatic embryos that produced green plants having a normal phenotype. Based on our observations with somatic embryogeny systems for various crops, conversion was defined as including all of the following:

- Germination (radicle elongation)
- Development of a vigorous root system
- Growth and development of the shoot meristem
- Production of at least two true leaves
- A direct shoot-to-root connection
- Absence of hypocotyl swelling
- Minimization of callus growth on the hypocotyl
- Production of a green plant with a normal phenotype.

In vitro conversion frequencies for alfalfa artificial seeds were increased from less than 1% to more than 60% for random-picked embryos by improving the embryogeny protocol and by focusing research efforts on embryo quality using conversion as a screening assay. A conversion frequency of 7% was achieved for alfalfa in soilless mix in the greenhouse. The lower conversion of in vivo planted embryos compared with in vitro indicated that, although a portion of the embryos were of sufficient quality to achieve greenhouse conversion, most had not reached the proper developmental stage. In later experiments, conversion frequencies of 20% were routinely obtained after the discovery that major nutrients (ammonium, calcium, magnesium, phosphate, and/or potassium) were required for conversion [34]. The conversion assay was critical for developing conditions and media that selected for uniform plant production.

B. Embryo Maturation

A cold maturation treatment was found to improve conversion of alfalfa somatic embryos, with an optimum period of 7–10 days at 4°C. No benefit was seen with a shorter period, and the embryos began to loose viability when stored longer [1].

Maltose was also effective for embryo maturation during the last stages prior to conversion. When conversion was done on a sucrose medium, the conversion frequency was the same whether the plants were scored after 4 or 6 wk. With maltose, however, there was an increase in the conversion frequency if the plants were scored after 6 wk [1]. This indicated that maltose served as a maturation factor for inducing further development of embryos that lagged behind the early, more mature embryos.

Morphological similarity between somatic embryos and zygotic embryos was realized at the time that the first somatic embryos were produced [38,39]. Since that time, patterns of somatic embryogeny have been well studied with the conclusion that remarkable homology exists between somatic and zygotic embryos [33]. Recent studies have also demonstrated biochemical similarities between zygotic embryos. Crouch [40] found both hypocotyl and microspore-derived embryos as well as zygotic embryos of Brassica napus L. to contain a 12S storage protein. Stuart et al. [41] and Stuart and Redenbaugh [42] found somatic and zygotic embryos of alfalfa to contain an 11S storage protein. Furthermore, alfalfa embryos, produced using a low 2,4-dichlorophenoxyacetic acid (2,4-D) induction, had increased levels of protein as well as a greater conversion frequency than embryos produced with higher 2,4-D [41]. Storage protein research was useful as both a confirmation that somatic embryos are similar to zygotic embryos and as a marker to assess maturation of the embryos.

Another potential embryo maturation marker is the use of isozymes. Starch gel electrophoresis was used to examine differences among various stages of zygotic embryogeny and between zygotic and somatic embryos of Brassica [34]. No differences were observed among immature zygotic embryos at different ages past pollination. However, 5 of the 15 isozymes used showed differential banding patterns between zygotic and somatic embryos. The differences appeared to be due to abnormal development of somatic embryos, which had low conversion frequencies and, consequently, were not of comparable quality to zygotic embryos.

C. Problems With Encapsulated Embryos

Conversion frequencies of encapsulated alfalfa somatic embryos used for early experiments were less than that for nonencapsulated embryos. It was discovered using seeds to study the encapsulation process that the root and shoot of the seedling had difficulty, at times, penetrating the outer surface of the capsule. This problem was resolved in a twofold fashion: 1) providing specific control over the alginate concentration (2% w/v) and limiting the complexing time to 30 min, and 2) producing somatic embryos that had higher conversion frequencies (greater quality). What became apparent was that the overall quality of the somatic embryos was critical for achieving high conversion frequencies and that encapsulation and coating systems, although problematic if not configured carefully, were not the limiting factors for development of artificial seeds [17].

Somatic embryos of other crops were encapsulated in sodium alginate and plants were recovered [1]. However, the frequency of embryo-to-plant conversion was low for rapid-cycling Brassica, carrot, cotton, lettuce, and corn because of the poor-quality embryos. Typical responses of poorer-quality embryos were callusing, hypertrophy of the hypocotyl, root growth without shoot growth, production of secondary embryos, production of fused cotyledons, and development of morphologically abnormal leaves. When poor-quality embryos were encapsulated, germination and conversion frequencies were generally reduced.

Embryogeny systems for most crops have not reached the level where encapsulation systems can adequately be assessed. Efforts to produce artificial seeds of such crops will be disappointing until greater emphasis is given to embryogeny. If in vitro conversion frequencies are low (less than 10%), then focus should be on embryogeny and not on encapsulation. The carrot system used for polyoxyethylene coating by Kitto and Janick [5,6] was not sufficient to assess the effect of the coating. Likewise, the early work of Redenbaugh et al. [30], who tested 27 gels for alfalfa embryo encapsulation, was lacking a satisfactory embryogeny system since only embryo survival could be measured. Clearly, embryo survival is not an adequate assay for development of artificial seeds. For a few species such as alfalfa (as it currently exists), the embryogeny protocols are sufficiently established so that high quality somatic embryos can be produced on a routine basis. As a result, encapsulation systems can now be modified, optimized, or newly developed using a standard somatic embryogeny protocol.

As modifications are made with the encapsulation system, it is anticipated that further improvements will be needed with somatic embryogeny. Improved synchrony of development will become critical as will control of contamination when the artificial seeds are planted under nonsterile environments. Most likely, encapsulation and embryogeny will be improved concurrently and tested one on the other as artificial seeds are developed to a commercial state. Still, the greatest research need is to produce somatic embryos that have achieved a level of developmental maturity that allows for seed-like conversion frequencies under greenhouse and field conditions.

V. COMMERCIALIZATION REQUIREMENTS FOR ARTIFICIAL SEEDS

A. Control of Somaclonal Variation

Tissue culture-derived variation is a major concern for development of artificial seeds because of the requirement for strict clonal fidelity. Variation has been reported for many crops, and seven case histories were recently presented [43]. Other reviews on the genetic and epigenetic effects of somaclonal variation have recently been published and will not be discussed here [44,45]. Orton [46] aptly summarized variation: "genetic instability in somatic tissues is ill understood and likely to stay that way for some time." Although the causes of somaclonal variation need not be well understood for commercialization of artificial seeds, control of the variation will be required. For one example discussed below, oil palm, variation was partially minimized by using an embryo/off-embryo production system in which the callus stage was essentially eliminated [47]. For other crops in which an extended callus/suspension culture stage is needed for scale-up, processes to control variation will be necessary.

B. Capsule Storage

Attention must be directed towards artificial seed storage. This will require maintaining a viable embryo for a discrete period of time without contamination. Ideally, artificial seeds should have a storage life as long as that for true seeds of any given crop. A minimum storage period of 6–12 mo or longer may be desirable for most crops, particularly where current seed technology allows for long shelf life. However, shorter storage periods may be tolerated if the value added for a specific crop is sufficiently great.

For hydrated alfalfa artificial seeds, the conversion frequencies of naked and encapsulated embryos (without storage) were equal. However, when

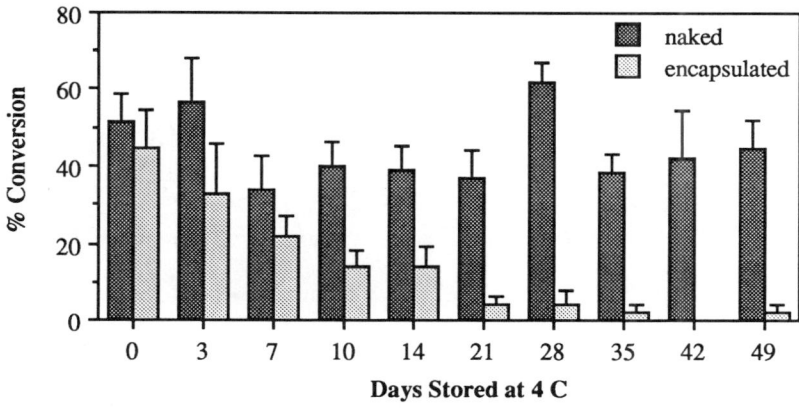

Fig. 3. Cold storage of alfalfa somatic embryos. (Reprinted from HortScience [30].)

stored, even at 4°C, the conversion frequency of encapsulated alfalfa somatic embryos declined sharply. The storage life for artificial seeds was found to be quite short using the current alginate protocol for alfalfa [30]. There was a significant decrease in subsequent conversion after 7 days; whereas the conversion frequencies of naked embryos was only slightly diminished over a 49-day period (Fig. 3). Preliminary results suggested that since the embryos were not quiescent but continued active respiration, the alginate capsule itself may be inhibiting respiration and thereby decreasing shelf life. However, the use of quiescent embryos does not, in itself, appear to overcome the short storage problem. Gray et al. [3] found that 7 days storage of desiccated, quiescent somatic embryos resulted in a severe drop in embryo conversion. These observations suggested that there was an interaction among various components of an artificial seed: the quality of the embryo, the stage of development (active respiration, quiescence), and the capsule matrix.

C. Bioreactor Production

A liquid flask system (Fig. 4) was developed for medium-scale production purposes for alfalfa based on the agar methods previously reported [36,48]. The agar method was modified to include a liquid induction stage of 3 days. The induced cells were collected on a 60-mesh stainless steel sieve and placed in liquid regeneration medium consisting of SH salts [35], 100 mM proline, and 25 mM ammonium. The embryos produced were collected on a 40 mesh sieve after 2 wk and resuspended in fresh, identical medium for a second 2-wk period. An additional improvement in embryo quality was to

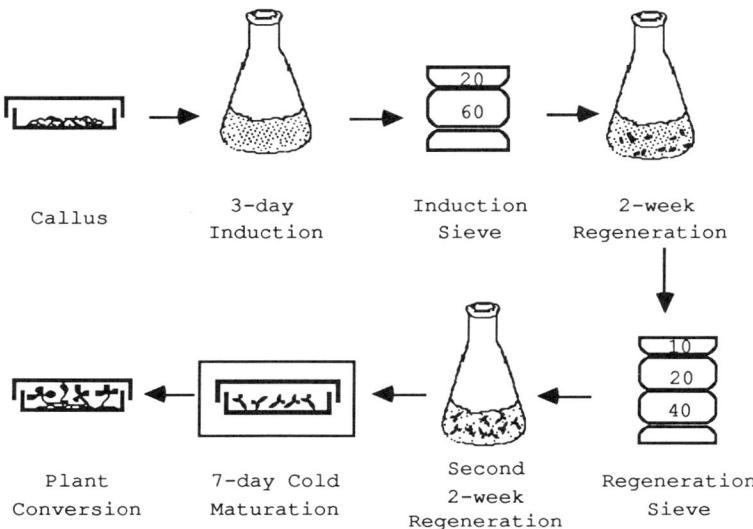

Fig. 4. Liquid production of alfalfa somatic embryos. (Reprinted courtesy of the VI International Congress of Plant Tissue and Cell Culture [1].)

wash the embryos with sterile water and store them at 4°C. A final modification was based on the discovery that maltose was effective for agar-based regeneration [37]. For conversion, 1.5% maltose was substituted for sucrose in half-strength SH medium. Embryos produced from the liquid flask cultures had a conversion frequency of 20–30% (using a random population greater than 1 mm length). The use of abscisic acid during the second regeneration stage had no consistent, positive effect.

For commercial production of somatic embryos, large-scale bioreactors will be required. Embryos will be produced using either batch cultures or harvested periodically from a continuous culture. Specialized vessels, such as spin bioreactors [26], may be used initially, followed by new bioreactors designed specifically for somatic embryo production.

D. Commercialization

Despite advances made in the past 30 years with the study of somatic embryos, only one species has reached pre-commercialization. Unilever Company began propagating and field testing oil palm in 1977 using an embryo/off-embryo propagation system. The process involves establishing a continuous embryo production cycle with little callus formation. Field tests to determine genetic stability of embryo-derived plants showed high clonal variation, equal to that of seedling controls. However, studies of variation in oil palm are complicated by high variation observed over time for individual plants [47]. For high per-unit value crops such as oil palm, the embryo/off-embryo method has potential. However, for most crops in which the per-unit seed value is low, such a labor-intensive system is not feasible.

VI. CONCLUSIONS

Somatic embryogenesis has been observed in a great many species to date, indicating that it may be possible to produce embryos in almost any desired crop. As more experience is gained with embryogeny, common features begin to emerge such as the discovery that somatic embryos could be produced using immature zygotic embryos from recalcitrant species like cereals [49] and conifers [50]. The use of the phenoxyacetic acids for inducing cells to develop into somatic embryos is a widespread requirement. Amino acids such as proline have been found useful for diverse species such as corn [51] and alfalfa [52]. It is anticipated that greater commonality will be found as our knowledge of somatic embryogeny increases.

Ten stages of commercialization of artificial seeds can be identified: 1) select candidate crops based on both technological and commercial potential, 2) optimize a somatic embryogenesis system, 3) optimize embryo maturation, 4) automate embryo production, 5) produce mature, synchronized embryos that have been developmentally arrested, 6) encapsulate embryos with necessary adjuvants, 7) coat encapsulated embryos with outer membrane, 8) optimize greenhouse and field conversion, 9) identify and control pest and disease problems that may be unique to artificial seeds, and 10) evaluate the artificial seed delivery system in terms of increased productivity for the specific crop [1].

Somatic embryos need to be produced using a mechanized, large-scale procedure, matured to a stage where germination will be at a high rate and frequency, and encapsulated in a suitable material. The encapsulated embryos will probably need to be coated to prevent capsule desiccation and allow for singulation during planting. Alternatively, the embryos may be desiccated and dry-coated. Then the artificial seeds will require a shelf life specific for the planting practices of the crop. The final assessment will be the harvest and yield.

ACKNOWLEDGMENTS

We thank Dr. Keith A. Walker for introducing us to the concept of artificial seeds and for his unwavering support over the last five years.

The hydrophobic coating for the alginate capsules was developed by Dr. Zoila Reyes (SRI International).

REFERENCES

1. Redenbaugh K, Viss P, Slade D, Fujii J: Scale-up: Artificial seeds. In Green C, Somers D, Hackett W, and Biesboer D (eds): "Plant Tissue and Cell Culture." New York: Alan R. Liss, Inc., 1987, p 473.
2. Kamada H: Artificial seed. In Tanaka R (ed): "Practical Technology on the Mass Production of Clonal Plants." Tokyo: CMC Publisher, 1985, p 48 (in Japanese).
3. Gray D, Conger B, Songstad D: Desiccated quiescent somatic embryos of orchardgrass for use as synthetic seeds. In Vitro Cell Dev Biol 23:29, 1987.
4. Hama I: Artificial seeds. Japanese Patent Application No. 40708/1986, Feb 27, 1986 (in Japanese).
5. Kitto S, Janick, J: Production of synthetic seeds by encapsulating asexual embryos of carrot. J Am Soc Hort Sci 110:277, 1985.
6. Kitto S, Janick, J: Hardening treatments increase survival of synthetically-coated asexual embryos of carrot. J Am Soc Hort Sci 110:283, 1985.
7. Lutz J, Wong J, Rowe J, Tricoli D, Lawrence R, Jr: Somatic embryogenesis for mass cloning of crop plants. In Henke R, Hughes K, Constantin M, Hollaender A (eds): "Tissue Culture in Forestry and Agriculture." New York: Plenum Press, 1985, p 105.
8. Murashige T: Plant cell and organ cultures as horticultural practices. Acta Hort 78:17, 1977.
9. Lawrence R, Jr: *In vitro* cloning systems. Environ Exp Bot 21:289, 1981.
10. Gray D: Introduction to the symposium. HortSci 22:796, 1987.
11. Drew R: The development of carrot (*Daucus carota* L.) embryoids (derived from cell suspension culture) into plantlets on a sugar-free basal medium. Hort Res 19:79, 1979.
12. White P: "The Cultivation of Animal and Plant Cells," 2nd edition. New York: Ronald Press, 1963.
13. Baker C: "Synchronization and Fluid Sowing of Carrot, *Daucus carota* Somatic Embryos." MS Thesis. Ann Arbor: University Microfilms, 1985.
14. Kitto S, Janick J: Polyox as an artificial seed coat for asexual embryos. HortSci 17:448, 1982 (abst).
15. Nitzsche W: One year storage of dried carrot callus. Z Pflanzenphysiol 100:269, 1980.
16. Redenbaugh K, Nichol J, Kossler M, Paasch B: Encapsulation of somatic embryos for artificial seed production. In Vitro 20:256, 1984 (abst).
17. Redenbaugh K, Paasch B, Nichol J, Kossler M, Viss P, Walker K: Somatic seeds: Encapsulation of asexual plant embryos. Bio/Technology 4:797, 1986.
18. Yamakawa K: Application of artificial seed and its potential. Agric. Chem. Today 29:68, 1985 (in Japanese).
19. Rogers M: Synthetic seed technology. Newsweek 102(22):111, 1983.
20. Walgate R: A germ of an idea. Nature 317:664, 1985.
21. Redenbaugh K: Analogs of botanic seed. U.S. Patent 4,562,663; 1986.
22. Nishimura S: Somatic seeds. In "4th Tokai Regional Meeting for Biotechnology," Nagoya, Japan 1986, p 12 (in Japanese).
23. Gupta P, Durzan D: Biotechnology of somatic polyembryogenesis and plantlet regeneration in loblolly pine. Bio/Technology 5:147, 1987.

24. Upadhyaya S, Gautz L, Garrett R: Retrofitting vegetable planters to seed gel encapsulated propagules. Appl Eng & Agri 3:211, 1987.
25. Durzan D: Progress and promise in forest genetics. In "Proc. 50th Anniv. Conf. 'Paper Science and Tech.—The Cutting Edge.'" Appleton, WI: The Institute of Paper Chemistry, 1980, p 31.
26. Styer D: Bioreactor technology for plant propagation. In Henke R, Hughes K, Constantin M, Hollaender A (eds): "Tissue Culture in Forestry and Agriculture." New York: Plenum Press, 1985, p 117.
27. de Vries J: Industry's view of seed quality and technological developments. Seed Quality Seminar (Royal Sluis), April 11–12, 1985, p 11.
28. Redenbaugh K: Delivery system for meristematic tissue. U.S. Patent 4,583,320; 1986.
29. Janick J, Kitto S: Process for encapsulating asexual plant embryos. U.S. Patent 4,615,141; 1986.
30. Redenbaugh K, Slade D, Viss P, Fujii J: Encapsulation of somatic embryos in synthetic seed coats. HortSci 22:803, 1987.
31. Redenbaugh K, Slade D, Viss P, Kossler M: Artificial seeds: Encapsulation of somatic embryos. In Valentine F (ed): "Colloquium on Progress and Prospects in Forest and Crop Biotechnology." New York: Springer-Verlag, 1988, in press.
32. Fujii J, Redenbaugh K, Walker K: Current status and future of artificial seed. In "Proc. Agriculture and Life Sciences in China," Third Annual Ideal Colloquium. Beltsville, MD: Inst. Int. Develop. and Education in Agriculture and Life Sciences, p 95.
33. Ammirato P: Patterns of development in culture. In Henke R, Hughes K, Constantin M, Hollaender A (eds): "Tissue Culture in Forestry and Agriculture." New York: Plenum Press 1985, p 9.
34. Redenbaugh K, Fujii J, Slade D, Viss P, Kossler M: Synthetic seeds—encapsulated somatic embryos. "78th Annual Meeting of the American Society of Agronomy," 1986, in press.
35. Schenk R, Hildebrandt A: Medium and techniques for induction and growth of mono-cotyledons and dicotyledons cell cultures. Can J Bot 50:199, 1972.
36. Stuart D, Strickland S: Somatic embryogenesis from cell cultures of *Medicago sativa* L. I. The role of amino acid additions to the regeneration medium. Plant Sci Lett 34:165, 1984.
37. Strickland S, McCall C, Nichol J, Stuart D: Enhanced somatic embryogenesis in *Medicago sativa* by addition of maltose and its higher homologs to the culture medium. In Somers D, Gengenbach B, Biesboer D, Hackett W, Green C (eds): "VI International Congress of Plant Tissue and Cell Culture Abstracts." Minneapolis: University of Minnesota, 1986, p 188.
38. Reinert J: Morphogenese und ihre Kontrolle an Gewebekulturen aus Carotten. Naturwis-senschaft 46:344, 1958.
39. Steward F, Mapes M, Mears K: Growth and organized development of cultured cells II. Organization in cultures grown from freely suspended cells. Am J Bot 45:705, 1958.
40. Crouch M: Non-zygotic embryos of *Brassica napus* L. contain embryo-specific storage proteins. Planta 156:520, 1982.
41. Stuart D, Nelsen J, Strickland S, Nichol J: Factors affecting developmental processes in alfalfa cell cultures. In Henke R, Hughes K, Constantin M, Hollaender A (eds): "Tissue Culture in Forestry and Agriculture." New York: Plenum Press, 1985, p 59.
42. Stuart D, Redenbaugh K: Use of somatic embryogenesis for the regeneration of plants. In LeBaron H, Mumma R, Honeycutt R, Duesing J (eds): "Biotechnology in Agricultural Chemistry." Washington, DC: American Chemical Society, 1987, p 87.

43. Vasil I (ed): "Cell Culture and Somatic Cell Genetics of Plants." Orlando: Academic Press, 1986, 657 pp.
44. D'Amato F: Cytogenetics of plant cell and tissue cultures and their regenerates. CRC Crit Rev Plant Sci 3:73, 1986.
45. Evans D, Sharp W: Somaclonal and gametoclonal variation. In Evans D, Sharp W, Ammirato P (eds): "Handbook of Plant Cell Culture." Vol 4. New York: Macmillan, 1986, p 97.
46. Orton T: Case histories of genetic variability *in vitro:* Celery. In Vasil I (ed): "Cell Culture and Somatic Cell Genetics of Plants." Orlando: Academic Press, 1986, p 345.
47. Brackpool A, Branton R, Blake J: Regeneration in palms. In Vasil I (ed): "Cell Culture and Somatic Cell Genetics of Plants." Vol 3. Orlando: Academic Press, 1986, p 207.
48. Walker K, Sato S: Morphogenesis in callus tissue of *Medicago sativa:* The role of ammonium ion in somatic embroyogenesis. Plant Cell Tissue Organ Culture 1:109, 1981.
49. Lu C, Vasil I: Somatic embryogenesis and plant regeneration from leaf tissues of *Panicum maximum* Jacq. Theor Appl Genet 59:275, 1981.
50. Hakman I, Fowke L, von Arnold S, Eriksson T: The development of somatic embryos in tissue culture initiated from immature embryos of *Picea abies* (Norway spruce). Plant Sci Lett 38:53, 1985.
51. Armstrong C, Green C: Establishment and maintenance of friable embryogenic maize callus and the involvement of L-proline. Planta 164:207, 1985.
52. Stuart D, Strickland S: Somatic embryogenesis from cell cultures of *Medicago sativa* L. II. The interaction of amino acids with ammonium. Plant Sci Lett 34:175, 1984.

Index

249

Contents of Previous Volumes